A MODERN APPROACH
TO CRITICAL PHENOMENA

The study of critical phenomena is one of the most exciting areas of modern physics. This book provides a thorough but economic introduction into the principles and techniques of the theory of critical phenomena and the renormalization group, from the perspective of modern condensed matter physics. Assuming basic knowledge of quantum and statistical mechanics, the book discusses phase transitions in magnets, superfluids, superconductors, and gauge field theories. Particular attention is given to modern topics such as gauge field fluctuations in superconductors, the Kosterlitz-Thouless transition, duality transformations, and quantum phase transitions, all of which are at the forefront of today's physics research.

A Modern Approach to Critical Phenomena contains numerous problems of varying degrees of difficulty, with solutions. These problems provide readers with a wealth of material to test their understanding of the subject. It is ideal for graduate students and more experienced researchers in the fields of condensed matter physics, statistical physics, and many-body physics.

IGOR HERBUT is Professor of Physics at Simon Fraser University in Burnaby, British Columbia. He has held visiting appointments at the Max Planck Institute, the Kavli Institute for Theoretical Physics, and the Tokyo Institute of Technology. Professor Herbut has authored a number of research papers on quantum phase transitions, disordered systems, gauge field theories, and high-temperature superconductivity.

T0215766

A MODERN APPROACH
TO CRITICAL PHENOMENA

IGOR HERBUT

Simon Fraser University

CAMBRIDGE
UNIVERSITY PRESS

CAMBRIDGE UNIVERSITY PRESS
Cambridge, New York, Melbourne, Madrid, Cape Town, Singapore,
São Paulo, Delhi, Dubai, Tokyo

Cambridge University Press
The Edinburgh Building, Cambridge CB2 8RU, UK

Published in the United States of America by Cambridge University Press, New York

www.cambridge.org
Information on this title: www.cambridge.org/9780521142380

First published 2007
This digitally printed version 2010

A catalogue record for this publication is available from the British Library

ISBN 978-0-521-85452-8 Hardback
ISBN 978-0-521-14238-0 Paperback

Dedicated to my parents, Divna and Fedor Herbut

Contents

Preface

It has been more than thirty years since the theory of universal behavior of matter near the points of continuous phase transitions was formulated. Since then the principles and the techniques of the theory of such "critical phenomena" have pervaded modern physics. The basic tenets of our understanding of phase transitions, the concepts of scaling and of the renormalization group, have been found to be useful well beyond their original domain, and today constitute some of our basic tools for thinking about systems with many interacting degrees of freedom. When applied to the original problem of continuous phase transitions in liquids, magnets, and superfluids, the theory is in remarkable agreement with measurements, and often even ahead of experiment in precision. For this reason alone the theory of critical phenomena would have to be considered a truly phenomenal physical theory, and ranked as one of the highest achievements of twentieth century physics.

The book before you originated in part from the courses on theory of phase transitions and renormalization group I taught to graduate students at Simon Fraser University. The students typically had a solid prior knowledge of statistical mechanics, and thus had some familiarity with the notions of phase transitions and of the mean-field theory, both being commonly taught nowadays as parts of a graduate course on the subject. In selecting the material and in gauging the technical level of the lectures I had in mind a student who not only wanted to become familiar with the basic concepts of the theory of critical phenomena, but also to learn how to actually use it to explain and compute. So I tried to provide the calculational details, particularly through solved problems, which would hopefully enable a motivated student to acquire what is today considered to be the standard working knowledge in the field, without having to take a separate course on field-theoretical techniques. The

ix

present book is an attempt to satisfy the perceived need for a graduate text that could accompany such a course.

The theme that runs through the book is the physics of the superfluid phase transition. There are several reasons for this. First, while historically it was the magnetic phase transitions for which the theory was first developed, the all-important notion of Ginzburg–Landau–Wilson theory is more naturally introduced for the system of interacting bosons. It is easy to then generalize the theory to other universality classes that include the more familiar Ising and Heisenberg magnetic phase transitions. Second, the superfluid order parameter allows the simplest topological defects, vortices, which are important in their own right and in fact play a crucial role at the superfluid phase transition. Finally, the superfluid critical point is experimentally the best quantitatively understood phase transition in nature, and as such provides the most stringent test for the theory.

A more experienced reader may notice the absence of so-called real-space renormalization on the pages that follow. While maybe more intuitive, the historically important method of real-space renormalization is much less systematic and general than Wilson's momentum-shell transformation, treated in detail here. If the reader is already familiar with real-space methods from a course on statistical mechanics, so much the better. But no such familiarity is in fact required. To draw an analogy with classical mechanics: while the concept of force is certainly important, one can almost completely dispose of it in favor of the Lagrangian or the Hamiltonian formulations. The general Ginzburg–Landau–Wilson field theory may be viewed as playing a somewhat similar role in the physics of continuous phase transitions.

The intended introductory level of the book notwithstanding, some of the chapters deal with a more advanced material. The selection criterion was that the subject, besides proven to be important and general, also had to be well established and relatively straightforward to discuss using the techniques already introduced elsewhere in the book. Chapter 4 deals with the issue of coupling of the order parameter to other soft modes, as exemplified by the Ginzburg–Landau theory of superconductors or the scalar electrodynamics. Chapter 7 deals with modern duality transformations which provide a precious non-perturbative perspective at some interesting phase transitions. These two chapters may be omitted at the first reading without consequences. Likewise, the sections in the remaining chapters marked with an asterisk represent more advanced material that, although in line with the rest of the book, may also be safely left for later times. On the other hand, some other important topics, like critical dynamics or phase transitions in disordered systems, have not been included. Although the selection of topics to some degree is certainly a matter of personal taste, the exclusion of these two may be partially justified by them not being on equally firm footing at the time of writing as the rest.

Conforming to my belief that physics is best learned by practising, the book contains numerous problems scattered throughout, all fully solved. Both the problems and their solutions either further illustrate some point in the main text, or provide complementary material interesting in its own right. Some problems are straightforward exercises, while others are more involved. Difficult, but often very instructive problems are again marked with an asterisk. The problem set represents an integral part of the book, and it is recommended that the reader goes through it as much as possible.

I am grateful to Matthew Case, Albert Curzon, Kamran Kaveh, Hidetoshi Nishimori, and Babak Seradjeh for reading parts of the manuscript and for their many useful suggestions for improvement. Of course, the responsibility for any remaining mistakes is solely mine. I am also grateful to Simon Fraser University for the sabbatical leave during which the manuscript was finalized, and to Masaki Oshikawa and the condensed matter theory group at the Tokyo Institute of Technology for their kind hospitality during that time. The last but not the least, I am thankful to my my wife Irena, and my children Leonard and Marlena, for tolerating long periods of my mental and physical absence.

1

Introduction

Phase transitions are defined, and the concepts of order parameter and spontaneously broken symmetry are discussed. Simple models for magnetic phase transitions are introduced, together with some experimental examples. Critical exponents and the notion of universality are defined, and the consequences of the scaling assumptions are derived.

1.1 Phase transitions and order parameters

It is a fact of everyday experience that matter in thermodynamic equilibrium exists in different macroscopic phases. Indeed, it is difficult to imagine life on Earth without all three phases of water. A typical sample of matter, for example, has the temperature–pressure phase diagram presented in Fig. 1.1: by changing either of the two parameters the system may be brought into a solid, liquid, or gas phase. The change of phase may be gradual or abrupt. In the latter case, the *phase transition* takes place at well defined values of the parameters that determine the phase boundary.

Phase transitions are defined as points in the parameter space where the thermodynamic potential becomes non-analytic. Such a non-analyticity can arise only in the thermodynamic limit, when the size of the system is assumed to be infinite. In a finite system the partition function of any system is a finite sum of analytic functions of its parameters, and is therefore always analytic. A sharp phase transition is thus a mathematical idealization, albeit one that describes the reality extremely well. Macroscopic systems typically contain $\sim 10^{23}$ degrees of freedom, and as such are very close to being in the thermodynamic limit. The phase boundaries in Fig. 1.1, for example, for this reason represent reproducible physical quantities.

1

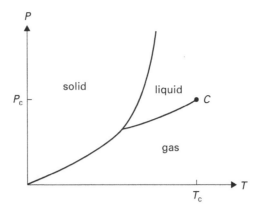

Figure 1.1 Temperature–pressure phase diagram of a typical piece of matter. All the phase transitions are discontinuous except at the critical point (*C*).

P. Ehrenfest gave an early classification of phase transitions according to the degree of the derivative of the thermodynamic potential with respect to the tuning parameter that is exhibiting a discontinuity. We will adhere to the simpler modern classification in which phase transitions may be either continuous (second order) or discontinuous (first order). A continuous phase transition is a change of phase of a macroscopic system in equilibrium not accompanied by latent heat. Phase transitions that do involve latent heat, like freezing of water for example, will be called first-order, or discontinuous. Non-analytic properties of systems near a continuous phase transition are called *critical phenomena*, and will be the main subject of this book. The point in the phase diagram where a continuous phase transition takes place is called a *critical point*.[1]

As the first step towards the understanding of phase transitions, it is useful to define a physical quantity that would clearly distinguish between different phases. Such an observable will be called the *order parameter*. The order parameter for a given phase transition may not be unique, and its choice is often dictated by its utility. For example, the liquid and gas phases may be distinguished by their average density. The liquid and solid phases differ in their densities as well, but obviously a more fundamental difference is that the density is uniform in the liquid and spatially periodic in the solid. The order

---

[1] Not all the quantities must change continuously near a critical point; in rare instances some may actually be discontinuous, like the superfluid density at the Kosterlitz–Thouless transition, discussed in Chapter 6. Also, there are a few examples of a weak first-order transition preempting a critical point, as in the type-I superconductors, which will also be considered here as critical phenomena in Chapter 4.

Table 1.1 *Examples of phase transitions and the corresponding order-parameters.*

System	Phase transition	Order parameter
H_2O, ^4He, Fe	liquid–solid	shear modulus
Xe, Ne, N_2, H_2O	liquid–gas	density difference
Fe, Ni	ferromagnet–paramagnet	magnetization
$RbMnF_2$, La_2CuO_4	antiferromagnet–paramagnet	staggered magnetization
^4He, ^3He	superfluid–normal liquid	superfluid density
Al, Pb, $YBa_2Cu_3O_{6.97}$	superconductor–metal	superfluid density
Li, Rb, H	Bose–Einstein condensation	condensate

parameter for the solid may therefore be defined as the Fourier transform of the density at some characteristic wavevector, so that it would be finite in the solid and zero in the liquid phase. Yet another choice would be the resistance to shear deformation, called the shear modulus, which is also finite only in the solid phase. This, as it turns out, is a more general choice, since in two dimensions a solid has a finite shear modulus while being of perfectly uniform density.

Many different phase transitions occur in nature. A small sample together with the appropriate order parameters is presented in Table 1.1. Some phase transitions are familiar from everyday life, while others must appear rather exotic. Nevertheless, there exists a coherent theoretical framework for detailed understanding of a variety of phase transitions. This is the subject of the present book.

1.2 Models: Ising, XY, Heisenberg

Before turning to the general theory of critical phenomena, it is useful to consider a specific model that actually exhibits a phase transition. Consider the partition function

$$Z = \sum_{\{s_i = \pm 1, i = 1, \ldots, N\}} e^{-\frac{E}{k_B T}}, \tag{1.1}$$

with the energy of a configuration $\{s_1, s_2, \ldots, s_N\}$ defined as

$$E = -J \sum_{\langle i,j \rangle} s_i s_j - H \sum_{i=1}^{N} s_i. \tag{1.2}$$

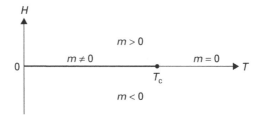

Figure 1.2 Temperature–magnetic field phase diagram of a magnetic system exhibiting a paramagnet–ferromagnet phase transition. The critical point at $T = T_c$ and $H = 0$ separates the ferromagnetic ($m \neq 0$) and the paramagnetic ($m = 0$) phases.

The discrete variables $s_i = \pm 1$ are defined on sites of a quadratic lattice, and may be understood as elementary magnetic dipoles that could point in one of the two directions. H then plays the role of an external magnetic field. The coupling $J > 0$ favors the pairs of neighboring dipoles, denoted by the symbol $\langle i, j \rangle$, to point in the same direction. At low temperatures ($T \ll J$), even if $H = 0$, one may expect all the dipoles to point in the same direction, while at high temperatures ($T \gg J$) the interaction between the dipoles becomes negligible, and consequently they will be completely randomly arranged. If we define the magnetization per dipole

$$m = \langle s_j \rangle = \frac{1}{Z} \sum_{\{s_i = \pm 1, i = 1, \dots, N\}} s_j e^{-\frac{E}{k_B T}} \tag{1.3}$$

as the order parameter, one may expect a phase transition between the ferromagnetic phase $m \neq 0$ and the paramagnetic phase $m = 0$, at $H = 0$ and the temperature $T = T_c$. For $H \neq 0$, in contrast, on average there will always be more dipoles pointing in the direction of the field, and magnetization will therefore be finite at all temperatures. The phase diagram of a ferromagnet is thus as given schematically at Fig. 1.2. The partition function in Eq. (1.1) was proposed by W. Lenz as the simplest model of ferromagnetism. The model may also be used to describe the structural order–disorder transition in binary alloys.

The "Ising model" in Eq. (1.1) can be solved exactly in one and two dimensions, as done by E. Ising and L. Onsager, respectively. The solution in one dimension is simple, but the magnetization turns out to vanish at all finite temperatures, and there is actually no phase transition (see Problems 1.1 and 1.2). The reason for this is easy to understand. Assume that an infinite one-dimensional array has all the dipoles pointing in the same direction. To flip half of the dipoles costs the energy $\Delta E = 2J$, since only one of a pair of nearest neighbors needs to point in opposite directions. This pair, on the other hand, can be chosen in N different ways, where $N \gg 1$ is the size of the system. The increase in entropy in the state with half of the dipoles flipped is therefore $\Delta S = k_B T \ln N$. The free energy $F = E - S$ of a large system may therefore always be lowered by flipping half of the dipoles, even at an infinitesimal temperature. The assumed ordered state is therefore unstable, and the equilibrium magnetization will consequently vanish at $T > 0$. Finite magnetization is thus possible only exactly at $T = 0$.

The above argument also suggests that the result of the competition between the energy and the entropy may be different in higher dimensions. Indeed, a more elaborate reasoning presented in Problem 1.3 was used by R. Peierls to argue that the state with spontaneous magnetization at finite temperatures should be possible in two dimensions. L. Onsager later succeeded in finding the exact solution and showed that there is a continuous phase transition in two dimensions at $k_B T_c / J = 2.269$. Furthermore, the spontaneous magnetization near and below T_c is $m \sim (T_c - T)^{1/8}$, and the specific heat diverges as $C \sim -\ln |T_c - T|$. The exact solution of the Ising model in two dimensions was the first demonstration that the equilibrium statistical mechanics can in principle lead to phase transitions. The solution, however, is of considerable complexity and not readily generalizable to other cases, so we will not describe it further here.

The Ising model in three dimensions has not been exactly solved at the time of writing. Nevertheless, we know with certainty that it does have a critical point, and even its quantitative characteristics are known with great accuracy. This comes from the systematic application of the theory exposed in the following chapters, as well as from numerical computations. The behavior of the magnetization and the specific heat, however, are different from the two-dimensional case. As may be suspected from the example of the Ising model, the system's *dimensionality* will in general play a crucial role in its critical behavior. Today there exist other exactly solvable models, which are often used as the testing ground for the general theory.

A notable feature of the Ising model is its global symmetry at $H = 0$ under the transformation $s_i \rightarrow -s_i$ at all sites. We will call this the Z_2, or the Ising, symmetry. This symmetry is obviously also present in the paramagnetic phase with $m = 0$, since any two configurations that have all the dipoles reversed enter the partition function with equal weight. In the ferromagnetic phase, however, the magnetization points in a definite direction, and the Ising symmetry is evidently broken. In the absence of an external magnetic field there is nothing that explicitly breaks the symmetry in the Hamiltonian, yet, the symmetry becomes broken *spontaneously* in the ordered phase. The direction of magnetization depends then on the history of the system. Both directions are equally probable, but once the direction has been randomly selected it becomes extremely unlikely that a macroscopic number of dipoles will be overturned by thermal fluctuations, so that the magnetization could change its sign. This phenomenon, in different forms quite ubiquitous in nature, is known as *spontaneous symmetry breaking*.

If the partition function for the Ising model is calculated by summing over all configurations, magnetization will of course always vanish due to the Z_2 symmetry. To describe the ordered phase with positive magnetization we must therefore restrict the space of configurations over which the summation in the partition function is to be performed. This may be conveniently done by calculating magnetization in a finite external magnetic field first, and then by taking the limit of zero magnetic field *after* the thermodynamic limit has been taken. The ordered phase will then end up having a finite magnetization surviving the limit of zero field. This mathematical procedure should not be understood as describing what literally occurs in the system, but only as a way of obtaining the correct physical result within the formalism of equilibrium statistical mechanics.

The actual process by which the direction of broken symmetry is selected is rather different, and no weak true magnetic field is involved at all. Let us define the dynamics of the Ising model in real time by letting the system evolve through different configurations chosen randomly in accordance with the Boltzmann distribution. We also require that a change from one configuration to the other can involve only a finite number of dipoles. At high temperatures the system will then explore the whole space of configurations, which results in vanishing magnetization. As the temperature is lowered those previously rare configurations of larger differences in numbers of up and down dipoles become more probable, by being favored by the interaction. The "time" it takes for the system to evolve from a configuration with a large block of dipoles pointing up, for example, to its Z_2 symmetric configuration also takes a progressively longer time, since the evolution takes place by flipping only a finite number of dipoles at a time, and each of these steps becomes less likely at lower temperatures. In thermodynamic limit the "time" it would take then for a macroscopic number of spins pointing up to get overturned becomes exponentially large. When observed at time intervals of long but finite length the *macroscopic* system thus exhibits a finite magnetization. The direction of this magnetization, however, is obviously random and essentially determined by the first configuration with a large enough imbalance of up and down dipoles that occurs in the system's evolution after the temperature was lowered below T_c.

The Ising model in zero magnetic field is thus invariant under global transformations belonging to the simplest *discrete symmetry group* $Z_2 = \{1, -1\}$. It is straightforward to generalize it to higher discrete symmetries, like $Z_3 = \{1, e^{i2\pi/3}, e^{i4\pi/3}\}$, which defines the family of so-called "clock models". A more substantial generalization is to the case of *continuous symmetry*, by allowing the dipoles of fixed magnitude to point arbitrarily in the plane (the "XY model"), or in space (the "Heisenberg model"), while remaining coupled ferromagnetically:

$$Z = \int \prod_{i=1}^{N} (\delta(|\vec{s}_i| - 1) \mathrm{d}^D \vec{s}_i) e^{\frac{J}{k_B T} \sum_{\langle i,j \rangle} \vec{s}_i \cdot \vec{s}_j + \frac{\vec{H}}{k_B T} \cdot \sum_{i=1}^{N} \vec{s}_i}, \tag{1.4}$$

where $D = 2$ for the XY, and $D = 3$ for the Heisenberg model. The symmetry of the model, or equivalently, of the order parameter, will turn out to be another decisive factor in its critical behavior.

We said nothing so far about the physical origin of the interaction between dipoles that is responsible for magnetic ordering. It should not be understood literally as the dipole–dipole interaction between spins of electrons in a solid, which is several orders of magnitude too weak to yield the observed critical temperatures of $T_c \sim 10^3 K$. It is instead an *effective* interaction accounting for the purely quantum mechanical exchange effect, and proportional to the Coulomb repulsion between electrons. A detailed discussion can be found in most books on quantum mechanics. Depending on the nature of electronic wave functions the coupling J may be either positive or negative, leading to ferromagnetic or antiferromagnetic orderings. In the latter case the dipoles on the neighboring sites point in the opposite directions. None of these complications matter, however, for the critical behavior of magnets near T_c: both $J > 0$ and $J < 0$ lead to spontaneous breaking of the same rotational symmetry, and turn out to exhibit the same critical behavior. The quantum mechanical nature of the elementary dipoles may be shown also to be irrelevant near T_c. This is what allows one to consider the grossly simplified classical models that we introduced. The only real novelty comes from permitting the sign of the interaction J to be random from one of a pair of neighbors to the other. This situation arises in certain alloys, such as CuMn. The low-temperature phase of these so-called "spin-glasses" has the dipoles frozen in time, but pointing randomly in space. Even the definition of an order parameter in this case becomes a rather subtle matter. This, however, lies beyond the scope of the present book.

Problem 1.1 Compute the free energy of the Ising model in one dimension in zero magnetic field.

Solution Let us define the link variables $t_i = s_i s_{i+1} = \pm 1$, with $i = 1, 2, \ldots, N - 1$. Each configuration of link variables corresponds to a unique, up to an overall sign, configuration of the dipoles. The partition function can therefore be written as

$$Z = 2 \sum_{\{t_i = \pm 1, i = 1, \ldots, N-1\}} e^{\frac{J}{k_B T} \sum_{i=1}^{N-1} t_i} = 2 \left[2 \cosh \frac{J}{k_B T} \right]^{N-1},$$

since different link variables decouple and factorize the partition function. The free energy $F = -k_B T \ln Z$ in the thermodynamic limit $N \to \infty$ is therefore

$$F = -N k_B T \ln \left[2 \cosh \frac{J}{k_B T} \right],$$

and evidently an analytic function of temperature.

Problem 1.2 Find the magnetization in the Ising model in one dimension with periodic boundary condition.

Solution To compute the magnetization m per dipole one needs the free energy in the external magnetic field so that

$$m = -\frac{1}{N} \frac{\partial F}{\partial H}.$$

With periodic boundary condition $s_N = s_1$ the partition function becomes

$$Z = \sum_{\{s_i = \pm 1, i = 1, \ldots, N-1\}} e^{\frac{J}{k_B T} \sum_{i=1}^{N-1} s_i s_{i+1} + \frac{H}{k_B T} \sum_{i=1}^{N-1} s_i} = \text{Tr}[\hat{M}^{N-1}],$$

where $\hat{M}$ is a 2×2 matrix with the matrix elements

$$M_{ss'} = e^{\frac{J}{k_B T} s s' + \frac{H}{2 k_B T} (s + s')}.$$

The partition function is therefore $Z = \lambda_+^{N-1} + \lambda_-^{N-1}$, where

$$\lambda_\pm = e^{\frac{J}{k_B T}} \left(\cosh \frac{H}{k_B T} \pm \left(\sinh^2 \frac{H}{k_B T} + e^{-\frac{4J}{k_B T}} \right)^{1/2} \right)$$

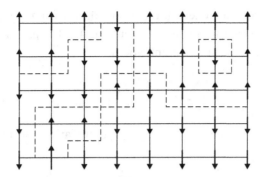

Figure 1.3 An example of a configuration of boundaries dividing the regions of positive and negative dipoles.

are the eigenvalues of the matrix $\hat{M}$. Since $\lambda_+ > \lambda_-$, in the thermodynamic limit $F = -Nk_BT \ln \lambda_+$. When $H = 0$ the free energy reduces to the one calculated in the previous problem. Finally, in the limit $H \rightarrow 0$ one finds $m = 0$ for all $T \neq 0$, and $m = 1$ at $T = 0$.

Problem 1.3* Formulate a qualitative argument in favor of the Ising model in two dimensions having $T_c > 0$.

Solution Any configuration can be described by drawing boundaries separating the regions of positive from negative dipoles, as for instance in Fig. 1.3. The boundary determines the configuration uniquely, up to an overall sign reversal. Measured from the energy of the perfectly ordered configuration, the energy is then $E = 2LJ$, where L is the total length of all boundaries, in units of lattice spacing. The partition function may then be written as

$$Z = 2 \sum_{L=0}^{\infty} G_L e^{-\frac{2JL}{k_BT}},$$

where G_L is the number of ways in which we can draw boundaries of total length L. In doing this, however, one must obey the following rules: (a) no boundary may pass the same segment more than once, (b) no two boundaries can overlap, (c) each boundary is either closed, or starts and ends on the edges of the system, and (d) if two boundaries intersect, one can draw the line for each using either branch beyond the intersection, with all the alternatives counted only once.

Obviously, these conditions make the exact computation of the number G_L a major problem. A qualitative estimate may be obtained by ignoring them.

Without any restrictions, one can draw boundaries from any point, and the number of ways to have it of length L would be 4^L. The partition function for the boundaries from the single point would be

$$Z' = \sum_{L=0}^{\infty} 4^L e^{-\frac{2JL}{k_B T}} = \frac{1}{1 - 4e^{-\frac{2J}{k_B T}}},$$

assuming $4e^{-\frac{2J}{k_B T}} < 1$. The average length of boundaries starting from a single point is then

$$\langle L \rangle = \frac{\sum_{L=0}^{\infty} L 4^L e^{-\frac{2JL}{k_B T}}}{Z'} = \frac{4e^{-\frac{2J}{k_B T}}}{1 - 4e^{-\frac{2J}{k_B T}}}.$$

How many overturned dipoles will these boundaries enclose? Here we may neglect the open boundaries that start and terminate at the edges, since their number is $\sim N^{1/2}$, whereas the number of closed boundaries is $\sim N$. A closed boundary can enclose at most $(L/4)^2$ dipoles, so the average number of overturned dipoles is less than

$$N \left(\frac{e^{-\frac{2J}{k_B T}}}{1 - 4e^{-\frac{2J}{k_B T}}} \right)^2.$$

When this number is smaller than $N/2$ the system will have a finite magnetization. This is the case for $k_B T < 1.184 J$. Since the argument overestimates the effect of thermal fluctuations we may expect that

$$k_B T_c > 1.18 J$$

for the Ising model in two dimensions. This is in accord with the exact result quoted in the text. With few modifications the argument may in fact be made quite rigorous and then used to prove the existence of a phase transition in the two-dimensional Ising model.

1.3 Universality and critical exponents

It must appear rather bizarre that physicists would devote their time to the study of continuous phase transitions. After all, most changes of phases in nature are in fact discontinuous. Take, for example, the phase diagram in Fig. 1.1: all the lines represent first-order phase transitions accompanied by latent heat, and it is only at the critical point C that the transition between the liquid and the gas phase is actually continuous. So why is that single point interesting? The reason is, at least in part, that some properties of the system near the critical point appear to be the same for completely different physical systems. This means, for example, that the specific heat near the liquid–gas critical point may behave the same way as the specific heat near the paramagnet–ferromagnet phase transition in an otherwise entirely different, magnetic system. Some macroscopic properties of a system near a continuous phase transition thus appear to be rather independent of the microscopic interactions between particles. As already hinted and will be discussed in greater detail later, they turn out to depend only on some broad characteristics of the system, like its dimensionality, symmetry, and presence or absence of sufficiently long-ranged interactions. Of course, without the microscopic interactions there would be no phase transitions at all, so one is faced with a somewhat paradoxical situation. The phenomenon of different systems exhibiting the same critical behavior is called the *universality*, and its explanation required a conceptually new approach to problems in statistical physics: the theory of the renormalization group. The renormalization group is devised to deal with the systems that exhibit fluctuations on a wide range of length scales, which we will see is what occurs near a critical point. Besides explaining the universality, the renormalization group also provides a general strategy for dealing with problems with many degrees of freedom, which turns out to be useful in other areas of physics. The concept of renormalization is crucial in the present theory of elementary particles, for example. Many other physical phenomena that appear self-similar on different length scales, like the turbulence in fluids, the distribution of earthquakes, or configurations of polymers, may also be studied by related methods. Finally, when augmented with some clever mathematical tricks, the theory of renormalization group may also be turned into a systematic calculational tool of the universal quantities.

As an example of the critical point, consider first the liquid–gas phase transition at $p = p_c$. If we define the *reduced temperature* $t = (T - T_c)/T_c$, the specific heat at fixed volume in the thermodynamic limit near the critical point behaves as the power-law

$$C_V = C_\pm |t|^{-\alpha},\tag{1.5}$$

where C_+ (C_-) refers to $t > 0$ ($t < 0$). Similarly, the compressibility at fixed temperature $K_T = (1/\rho)\partial\rho/\partial p$, with $\rho = N/V$ being the particle density, behaves as

$$K_T = K_\pm |t|^{-\gamma}.\tag{1.6}$$

The difference between the gas and liquid densities, which we defined as the order parameter for the liquid–gas transition, along the coexistence curve near the critical point has the form

$$\rho_L - \rho_G = \rho_c(-t)^\beta. \tag{1.7}$$

where ρ_c is the density at the critical point. Furthermore, at the critical isotherm $T = T_c$, near $p = p_c$,

$$\frac{p - p_c}{p_c} = \left|\frac{\rho_L - \rho_G}{\rho_c}\right|^\delta. \tag{1.8}$$

The powers $\alpha, \beta, \gamma, \delta$ are some non-trivial, and, as will be shown, not completely independent real numbers. They are *universal*, i.e. the same for a whole class of various phase transitions. An old and famous example of the universality of the liquid–gas transition in different fluids is provided by the plot in Fig. 1.4. We will call these numbers *critical exponents*, and one task of the theory of critical phenomena will be to explain why and to what degree they are universal, and to find a way to compute them. Note that the critical exponents α and γ may in principle be defined both below and above the critical point. While one would expect that the exponents describing the power-law for the divergence of the specific heat, for example, could be different above and below T_c, the theory predicts that the two values are in fact the same.

Critical exponents are not the only universal quantities. There are others, like scaling functions and universal amplitude ratios, like C_+/C_- and K_+/K_-. The exponents are, however, often the most directly measurable quantities, and their non-trivial and nearly universal experimental values provided the main motivation for the theory of critical phenomena.

1.4 Scaling of free energy

The power-laws near the critical point can be derived from the assumption of *scaling*. To be specific, consider a magnetic system near the paramagnet–ferromagnet transition at Fig. 1.2. For simplicity we will assume that magnetization has a preferred axis and thus may be assumed to be a scalar, i.e. a "uniaxial ferromagnet". With this restriction the system has the same symmetry as the Ising model, and we will say it belongs to the "Ising universality class". We assume that the Gibbs free energy per unit volume, which in principle is a function of the temperature T and the external magnetic field H,

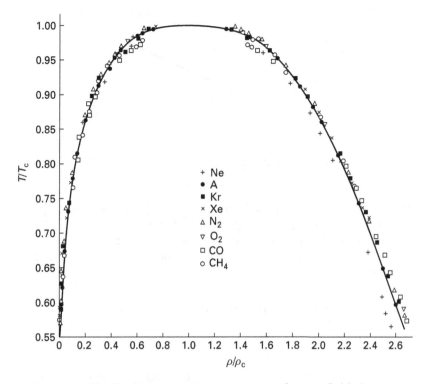

Figure 1.4 The liquid–gas coexistence curves of many fluids become identical when the axis are rescaled with the critical values of density and temperature (see Problem 1.5). The curve is, however, much closer to the cubic (shown here) than to the quadratic dependence, which implies $\beta \approx 1/3$. (Reprinted with permission from E. A. Guggenheim, *Journal of Chemical Physics* **13**, 253 (1945). Copyright 1945, American Institute of Physics.)

near the critical point at $T = T_c$ and $H = 0$, can be written as

$$f(T, H) = |t|^{1/w} \Psi_{\pm}\left(\frac{H}{|t|^{u/w}}\right),\tag{1.9}$$

where $\Psi_+(z)$ for $t > 0$, and $\Psi_-(z)$ for $t < 0$, are two different functions of a single variable. Note that even before we specify the forms of $\Psi_{\pm}(z)$, or the numbers u and w, the scaling assumption in Eq. (1.9) severely restricts the possible form of $f(T, H)$. The point is that once the assumption is made, the power laws for the magnetization, specific heat, and susceptibility, analogous to those for the liquid–gas transition in Eqs. (1.5)–(1.8) immediately follow, together with two equations satisfied by the four critical exponents.

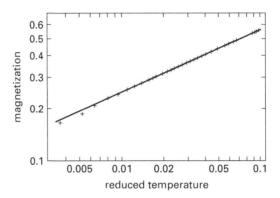

Figure 1.5 Magnetization vs. reduced temperature in nickel on a log–
log plot. The slope of the straight line yields the exponent $\beta = 0.358$.
(Reprinted with permission from J. D. Cohen and T. R. Carver, *Physical
Review B* **15**, 5350 (1977). Copyright 1977 by the American Physical
Society.)

Consider first the magnetization per unit volume at $H = 0$,

$$m = -\frac{\partial f}{\partial H}\Big|_{H=0} = -|t|^{\frac{1-u}{w}}\Psi'_{\pm}(0). \tag{1.10}$$

Since for $t > 0$ the system is a paramagnet and $m \equiv 0$, it must be that
$\Psi'_{+}(0) = 0$. Magnetization is the order parameter for the paramagnet–
ferromagnet transition, analogous to the density difference in the case of
the liquid–gas transition. This suggests we define the exponent β by analogy
to Eq. (1.7) as

$$m \propto (-t)^{\beta}, \tag{1.11}$$

for $t < 0$. A good example of such a power-law dependence of magnetization
on the reduced temperature in nickel is given in Fig. 1.5. Comparing with
Eq. (1.10), we see that

$$\beta = \frac{1 - u}{w}. \tag{1.12}$$

For the uniform magnetic susceptibility we similarly find that near T_c

$$\chi = -\frac{\partial^2 f}{\partial H^2}\Big|_{H=0} = -|t|^{\frac{1-2u}{w}}\Psi''_{\pm}(0). \tag{1.13}$$

Since the susceptibility is the derivative of order parameter with respect to
the magnetic field, which is its thermodynamically conjugate variable, it is
analogous to the compressibility in Eq. (1.6). So we define the exponent γ for

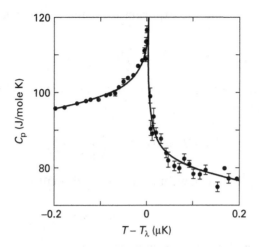

Figure 1.6 Specific heat at constant pressure near the superfluid transition in ^{4}He. The best fit to the power-law $C_P = B + A_\pm |t|^{-\alpha}/\alpha$ yields the exponent $\alpha = -0.0127 \pm 0.0003$. The negative value of the exponent means that the specific heat at $T = T_c$ in this particular instance is actually finite, and its smallness that the temperature dependence is very close to being logarithmic. The minimal value of reduced temperature in this measurement is $t \approx 10^{-8}$! (Reprinted with permission from J. A. Lipa, J. A. Nissen, D. A. Stricker, D. R. Swanson, and T. C. P. Chui, *Physical Review B* **68**, 174518 (2003). Copyright 2003 by the American Physical Society.)

the ferromagnet–paramagnet transition as

$$\chi \propto |t|^{-\gamma}, \tag{1.14}$$

and the scaling assumption implies that

$$\gamma = \frac{2u - 1}{w}. \tag{1.15}$$

The divergence in the specific heat may be similarly found:

$$C_V = -T\frac{\partial^2 f}{\partial T^2}\Big|_{H=0} = -\frac{\Psi_\pm(0)}{T_c}\frac{1}{w}\left(\frac{1}{w} - 1\right)t^{\frac{1}{w}-2}, \tag{1.16}$$

so that the exponent α is given by

$$\alpha = 2 - \frac{1}{w}. \tag{1.17}$$

The most precisely measured value of any critical exponent is the value of α at the superfluid transition in helium (Fig. 1.6).

Finally, we may define the exponent δ for the paramagnet–ferromagnet transition as

$$m \propto H^{1/\delta}, \tag{1.18}$$

at the critical point $t = 0$, and as $H \to 0$. Rewriting the scaling assumption in Eq. (1.9) slightly as

$$f(T, H) = H^{1/u} \tilde{\Psi}_{\pm} \left(\frac{H}{|t|^{u/w}} \right), \tag{1.19}$$

where $\tilde{\Psi}_{\pm}(z) = z^{-1/u} \Psi_{\pm}(z)$, we find at $t = 0$ and small H

$$m = -\frac{1}{u} H^{\frac{1}{u}-1} \tilde{\Psi}_{\pm}(\infty). \tag{1.20}$$

The exponent δ therefore depends only on u, and

$$\delta = \frac{u}{1 - u}. \tag{1.21}$$

Only two out of four defined critical exponents are therefore independent. In particular, eliminating u and w we find that

$$\alpha + 2\beta + \gamma = 2, \tag{1.22}$$

and

$$\alpha + \beta(\delta + 1) = 2. \tag{1.23}$$

The last two equations are often called Rushbrooke's and Griffiths' scaling laws. Thermodynamics alone requires that both be satisfied only as inequalities,[2] and the experimental indication that they in fact may be equalities motivated the introduction of the scaling hypothesis.

The scaling form of the free energy in Eq. (1.9) guarantees that the critical exponents are the same below and above the critical point. The same is not true for the whole scaling function, and we saw that in fact two different forms are needed for $T > T_c$ and $T < T_c$. Besides the critical exponents, which we already introduced as universal quantities, there are other quantities contained in the scaling functions $\Psi_{\pm}(z)$ that are also universal. For example, the ratio of the specific heats above and below T_c,

$$\frac{C_V(t \to 0+)}{C_V(t \to 0-)} = \frac{\Psi_+(0)}{\Psi_-(0)}, \tag{1.24}$$

---

[2] R. B. Griffiths, *Physical Review Letters* **14**, 623 (1965).

is dimensionless, and also universal. This is an example of a universal *amplitude ratio*. An analogous universal ratio may also be constructed for the susceptibility, for example. The universality of all these quantities is the consequence of the assumed scaling form of the free energy.

The concept of scaling therefore rationalizes the appearance of power-laws near the critical point, and yields the experimentally correct relations between the critical exponents. The questions still remain, however, why the scaling assumption should hold, what determines the scaling functions, and how one would compute them. The simple Curie–Weiss mean-field theory for magnetic systems, or the van der Waals theory of the liquid–gas phase transition, for example, gives $\beta = 1/2$ and $\alpha = 0$ (Problems 1.4 and 1.5), while in reality $\beta \approx 1/3$, and α is small but finite. In spite of the numerical difference between the mean-field and the experimental values of the exponents being rather small, a novel strategy is required to improve upon the mean-field theory. Before we come to that, however, two more exponents need to be introduced.

Problem 1.4 Find the critical exponents and the amplitude ratio for the magnetic susceptibility for the Ising model in the Curie–Weiss mean-field approximation.

Solution In the mean-field approximation each dipole is assumed to feel only the average local magnetic field due to other dipoles. This amounts to the replacement of the coupling term $J \sum_{i,j} s_i s_j$ in the Ising model with $J \sum_{i,j} \langle s_i \rangle s_j$. The average magnetization per spin $m = \langle s_i \rangle$ may then be determined self-consistently by

$$m = \frac{\sum_{s=\pm 1} s e^{\frac{(zmJ+H)s}{k_B T}}}{\sum_{s=\pm 1} e^{\frac{(zmJ+H)s}{k_B T}}},$$

where $z = 2d$ is the number of nearest neighbors on the quadratic lattice in d dimensions, and H the external magnetic field. So

$$m = \tanh\left(\frac{zmJ + H}{k_B T}\right),$$

and at $H = 0$, $m \neq 0$ for $T < T_c = zJ/k_B$, and $m = 0$ for $T > T_c$. Expanding near $m = 0$ and at $H = 0$, one finds near and below T_c

$$m = \frac{T}{T_c}\sqrt{3\left(1 - \frac{T}{T_c}\right)},$$

and thus $\beta = 1/2$. Right at $T = T_c$, for small H one similarly finds

$$m = \left(\frac{3H}{k_B T_c} \right)^{1/3},$$

and $\delta = 3$. Expanding the equation for magnetization above and near T_c around $m = 0$, it follows that

$$\chi(T > T_c) = \frac{dm}{dH} = \frac{1}{k_B(T - T_c)}.$$

Below and near T_c, on the other hand,

$$\chi(T < T_c) = \frac{1}{2k_B(T_c - T)}.$$

So the exponent $\gamma = 1$, both below and above T_c. The susceptibility ratio, however, is

$$\frac{\chi(T \to T_c^+)}{\chi(T \to T_c^-)} = 2.$$

Similarly, one finds that the specific heat has a discontinuity at T_c, and thus $\alpha = 0$.

Problem 1.5 Find the critical exponents at the liquid–gas transition in the non-ideal gas, described by the van der Waals equation of state $(p + (a/v^2))$ $(v - b) = k_B T$, where a and b are phenomenological parameters, and $v = V/N$ the volume per particle.

Solution From $p = -\partial F/\partial V$ at constant T, by demanding that for $a = b = 0$ one recovers the free energy of the ideal gas, we find

$$F = -Nk_B T \ln \left(1 - \frac{b}{v} \right) - \frac{aN}{v} + F_{ideal},$$

so that the heat capacity at a fixed volume is

$$C_V = -T \frac{\partial^2 F}{\partial T^2} = \frac{3Nk_B}{2},$$

implying $\alpha = 0$.

For $T < T_c$ there are three real solutions for v that satisfy the equation of state:

$$v^3 - \left(b + \frac{k_B T}{p} \right) v^2 + \frac{a}{p} v - \frac{ab}{p} = 0.$$

The three roots of this cubic equation merge into one as $T \rightarrow T_c^-$. At $T = T_c$ and $p = p_c$, therefore, the equation of state simplifies into $(v - v_c)^3 = 0$. When compared with the general equation this determines $v_c = 3b$, $p_c = a/(27b^2)$, and $k_B T_c = 8a/(27b)$. Rescaling the parameters as $p/p_c \rightarrow p$, $v/v_c \rightarrow v$, and $T/T_c \rightarrow T$, the van der Waals equation may be cast into the universal form

$$\left(p + \frac{3}{v^2} \right)(3v - 1) = 8T,$$

also known as the *law of corresponding states*. Near the critical point one may expand

$$p - 1 = 4(T - 1) - 6(T - 1)(v - 1) - \frac{3}{2}(v - 1)^3$$
$$+ O((T - 1)(v - 1)^2, (v - 1)^4)$$

and determine the actual liquid and gas densities from the Maxwell construction

$$\int_{u_L}^{u_G} u \, dp = \int_{u_L}^{u_G} u \left\{ -6(T - 1) - \frac{9}{2}u^2 \right\} du = 0,$$

where $u = v - 1$. Since the above equation should hold at any temperature close to and below the critical value, the interval of integration must be symmetric: $u_L = -u_G$. From the equation for pressure it then follows that

$$u_G \sim \sqrt{1 - T},$$

and $\beta = 1/2$. Similarly, right at $T = 1$,

$$p - 1 \sim -\frac{3}{2}(v - 1)^3$$

and thus $\delta = 3$. From $du/dp \sim 1/t$ it follows that $\gamma = 1$.

The law of corresponding states was the first example of the universality, and indeed has been observed in many liquid–gas transitions (Fig. 1.4). The exponents $\beta \approx 1/3$, however, instead of the mean-field value of $1/2$.

1.5 Correlations and hyperscaling

Define the two-point *correlation function* as

$$G(r, t) = \langle (m(\vec{r}) - m)(m(0) - m) \rangle = \langle m(\vec{r})m(0) \rangle - m^2, \qquad (1.25)$$

where $m(\vec{r})$ stands for the local value of magnetization, and $\langle \ldots \rangle$ denotes the ensemble average. An analogous correlation function in terms of the particle densities may be constructed for the liquid–gas system, and with some care for magnetic systems outside the Ising universality class. Assume then the following scaling form for the correlation function:

$$G(r, t) = \frac{\Phi_\pm(r/\xi(t))}{r^{d-2+\eta}}, \tag{1.26}$$

where $\xi(t) \propto |t|^{-\nu}$ is the *correlation length*, which should be diverging near the critical point. d is the dimensionality of the system, and ν and η two new critical exponents. Again, we expect two different scaling functions above and below T_c.

Using the general theory of linear response we may relate the exponents ν and η to the exponent γ. Defining the non-local magnetic susceptibility as

$$\chi(\vec{r} - \vec{r}') = \frac{\partial m(\vec{r})}{\partial H(\vec{r}')}|_{H=0}, \tag{1.27}$$

the general theory of linear response implies that

$$\chi(\vec{q}) = G(\vec{q}). \tag{1.28}$$

The uniform ($q = 0$) susceptibility χ introduced in Eq. (1.13) is therefore

$$\chi = \chi(\vec{q} = 0) = \int d^d\vec{r} G(\vec{r}, t). \tag{1.29}$$

Using the scaling assumption in Eq. (1.26), it then follows that

$$\chi = \text{const} \times \xi^{2-\eta}, \tag{1.30}$$

where the constant is given by

$$\text{const} = \int d^d\vec{z} \frac{\Phi_\pm(z)}{z^{d-2+\eta}}. \tag{1.31}$$

Since the values of local magnetization at two distant points should be uncorrelated for $T > T_c$, $\Phi_+(z)$ should be exponentially small for large arguments. Although $m \neq 0$ for $T < T_c$, the deviations from the finite magnetization are also uncorrelated at large distances, and the same is true for $\Phi_-(z)$. The integral in the last equation is therefore finite in both cases. Comparing with the definition of the exponent γ we see that

$$\gamma = \nu(2 - \eta), \tag{1.32}$$

which is also known as Fisher's scaling law.

Table 1.2 *Measured values of critical exponents in different systems,*
belonging to the Ising, XY, and Heisenberg universality classes.

	Xe	^{4}He	Fe	Ni
Universality class	Ising	XY	Heisenberg	Heisenberg
α	< 0.2	-0.0127 ± 0.0003	-0.03 ± 0.12	0.04 ± 0.12
β	0.35 ± 0.015		0.37 ± 0.01	0.358 ± 0.003
γ	1.3 ± 0.2		1.33 ± 0.015	1.33 ± 0.02
δ	4.3 ± 0.5		4.3 ± 0.1	4.29 ± 0.05
η	0.1 ± 0.1		0.07 ± 0.04	0.041 ± 0.01
ν	≈ 0.57	0.6705 ± 0.0006	0.69 ± 0.02	0.64 ± 0.1

Finally, assuming that the only relevant length scale near T_c is provided by the correlation length ξ, the free energy per unit volume is expected to scale as

$$f \propto \xi(t)^{-d}. \tag{1.33}$$

Differentiating twice with respect to temperature, one finds the specific heat

$$C \propto |t|^{\nu d-2}, \tag{1.34}$$

and thus

$$\alpha = 2 - \nu d. \tag{1.35}$$

The last relation is known as Josephson's scaling law. Unlike the remaining three relations between the critical exponents, Josephson's scaling law involves the dimensionality of the system. Since its derivation required an additional scaling assumption in Eq. (1.33), it is also often referred to as the *hyperscaling*. We will see that in contrast to the remaining three scaling laws the hyperscaling is not always satisfied.

To summarize, we have introduced six critical exponents to describe singular behavior of various thermodynamic functions and of the correlation function near the magnetic and liquid–gas critical points. We introduced the scaling ansatz for the free energy and the correlation function, and derived four equations (including hyperscaling) for the exponents, reducing the number of independent exponents to only two. Whenever the hyperscaling holds, we may then compute the correlation length exponent ν and the *anomalous dimension* η, for example, to determine all six critical exponents. Some examples of the values of critical exponents in different systems are given in Table 1.2.

Problem 1.6 Assuming that the Fourier transform of the correlation function has the form $G^{-1}(\vec{q}) = \vec{q}^2 + t$, for $t > 0$ and in three dimensions, find the correlation length exponent ν and the anomalous dimension η.

Solution Performing the Fourier transform we find

$$G(\vec{r}) = \int \frac{d^3\vec{q}}{(2\pi)^3} \frac{e^{i\vec{q}\cdot\vec{r}}}{q^2 + t} = \frac{e^{-|\vec{r}|\sqrt{t}}}{4\pi r},$$

and thus $\nu = 1/2$ and $\eta = 0$.

Problem 1.7 Combining the mean-field exponents from Problem 1.4 with the result of the previous problem, check the validity of the four scaling laws. In which dimension is the Josephson's scaling law satisfied?

Solution The first three scaling laws are all satisfied. Josephson's law is satisfied only in four dimensions.

2

Ginzburg–Landau–Wilson theory

The partition function for interacting bosons is derived as the coherent state path integral and then generalized to magnetic transitions. Phase transitions in the Ginzburg–Landau–Wilson theory for a fluctuating order parameter are discussed in Hartree's and Landau's approximations, and the fundamental limitation of perturbation theory near the critical point is exposed. The concept of upper critical dimension is introduced.

2.1 Partition function for interacting bosons

As a prototypical system with a continuous phase transition we will consider the system of interacting bosons. A well studied physical realization is provided by helium (^{4}He) with the pressure–temperature phase diagram as depicted in Fig. 2.1. Since the atoms of helium are light and interact via weak dipole–dipole interactions, due to quantum zero-point motion helium stays liquid down to the lowest temperatures, at not too high pressures. Instead of solidifying it suffers a continuous normal liquid–superfluid liquid transition at $T_c \approx 2K$, also called the λ-transition due to the characteristic form of the specific heat in Fig. 1.6. The λ-transition represents the best quantitatively understood critical point in nature. We have already quoted the specific heat exponent $\alpha = -0.0127 \pm 0.0003$, with the power-law behavior being observed over six decades of the reduced temperature! To achieve this accuracy the experiment had to be performed in the space shuttle so that the small variations in T_c along the height of the sample due to Earth's gravity would be minimized. At higher pressures ^{4}He eventually solidifies, with the superfluid–solid and the normal liquid–solid phase transitions being discontinuous, the former being so rather weakly.

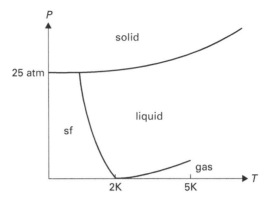

Figure 2.1 Phase diagram of helium (^{4}He).

Before we begin the study of the superfluid phase transition we will need a particular representation of the partition function for the system. Once this representation of the partition function describing the superfluid transition is derived it will prove possible to alter it only slightly and describe the critical behavior in the other universality classes, and in particular the magnetic transitions, as well. So let us recall the standard orthogonal basis of quantum mechanical many-body states for the bosonic system:

$$|n_{\alpha_1}, n_{\alpha_2}, \ldots, n_{\alpha_N}\rangle = \prod_{i=1}^{N} \frac{(\hat{a}_{\alpha_i}^{\dagger})^{n_{\alpha_i}}}{\sqrt{n_{\alpha_i}!}} |0\rangle, \tag{2.1}$$

where $\{\alpha_i\}$ label the states that form a basis in the single-particle Hilbert space of dimension N, and $|0\rangle$ is the vacuum. $\hat{a}_{\alpha_i}^{\dagger}$ and $\hat{a}_{\alpha_i}$ are the bosonic creation and annihilation operators that create or destroy a boson in the single-particle state $|\alpha_i\rangle$. They satisfy the standard commutation relations $[\hat{a}_{\alpha_i}, \hat{a}_{\alpha_j}^{\dagger}] = \delta_{\alpha_i, \alpha_j}$, with all creation (annihilation) operators commuting among themselves. Any many-particle state can then be written as a linear combination of the basis states

$$|\Phi\rangle = \sum_{n_{\alpha_1}=0}^{\infty} \cdots \sum_{n_{\alpha_N}=0}^{\infty} \Phi_{n_{\alpha_1}, \ldots, n_{\alpha_N}} |n_{\alpha_1}, \ldots, n_{\alpha_N}\rangle. \tag{2.2}$$

Define a *coherent state* as the common eigenstate of the annihilation operators:

$$\hat{a}_{\alpha_i} |\Phi\rangle = \Phi_{\alpha_i} |\Phi\rangle, \tag{2.3}$$

where $i = 1, 2, \ldots, N$, and the eigenvalues Φ_{α_i} are complex. For such a state,

$$\Phi_{n_{\alpha_1}, \ldots, n_{\alpha_N}} = \prod_{i=1}^{N} \frac{(\Phi_{\alpha_i})^{n_{\alpha_i}}}{\sqrt{n_{\alpha_i}!}}, \tag{2.4}$$

and therefore

$$|\Phi\rangle = \sum_{n_{\alpha_1}=0}^{\infty} \cdots \sum_{n_{\alpha_N}=0}^{\infty} \prod_i \frac{(\Phi_{\alpha_i} \hat{a}_{\alpha_i}^{\dagger})^{n_{\alpha_i}}}{n_{\alpha_i}!} |0\rangle = e^{\sum_{\alpha_i} \Phi_{\alpha_i} \hat{a}_{\alpha_i}^{\dagger}} |0\rangle. \tag{2.5}$$

The last expression for the coherent state can easily be seen to satisfy the definition in Eq. (2.3) by applying the annihilation operator and by using the commutation relation $[\hat{a}_{\alpha_i}, (\hat{a}_{\alpha_j}^{\dagger})^n] = n(\hat{a}_{\alpha_i}^{\dagger})^{n-1} \delta_{\alpha_i, \alpha_j}$. Note that the coherent state is then a linear combination of states with different numbers of particles.

Similarly, one can define the bra version of a coherent state

$$\langle\Phi| = \langle 0|e^{\sum_{\alpha_i} \Phi_{\alpha_i}^{*} \hat{a}_{\alpha_i}}, \tag{2.6}$$

so that $\langle\Phi|\hat{a}_{\alpha_i}^{\dagger} = \langle\Phi|\Phi_{\alpha_i}^{*}$. Then

$$\hat{a}_{\alpha_i}^{\dagger} |\Phi\rangle = \frac{\partial}{\partial\Phi_{\alpha_i}} |\Phi\rangle. \tag{2.7}$$

The overlap of two coherent states follows to be

$$\langle\Phi|\Phi'\rangle = e^{\sum_{\alpha_i} \Phi_{\alpha_i}^{*} \Phi_{\alpha_i}'}, \tag{2.8}$$

so that different coherent states are not orthogonal. Nevertheless, they form an overcomplete set, and one can write the resolution of unity as

$$1 = \int \prod_\alpha \frac{d\Phi_\alpha^* d\Phi_\alpha}{2\pi i} e^{-\sum_\alpha \Phi_\alpha^* \Phi_\alpha} |\Phi\rangle\langle\Phi|, \tag{2.9}$$

where the index i will be omitted hereafter for simplicity.

To prove Eq. (2.9), consider the one-dimensional single-particle Hilbert space, when $N = 1$. Then,

$$\int \frac{d\Phi^* d\Phi}{2\pi i} e^{-|\Phi|^2} |\Phi\rangle\langle\Phi|$$

$$= \int_0^\infty \frac{r\,dr}{\pi} \int_0^{2\pi} d\theta e^{-r^2} \sum_{m=0}^{\infty} \sum_{n=0}^{\infty} \frac{(re^{i\theta})^m}{\sqrt{m!}} |m\rangle \frac{(re^{-i\theta})^n}{\sqrt{n!}} \langle n|$$

$$= \sum_{n=0}^{\infty} \frac{1}{n!} \int_0^\infty dr 2r e^{-r^2} r^{2n} |n\rangle\langle n| = \sum_{n=0}^{\infty} |n\rangle\langle n| = 1. \tag{2.10}$$

The proof directly generalizes to $N > 1$.

Having the resolution of unity, we may proceed to derive the path-integral representation of the grand-canonical partition function,

$$Z = \mathrm{Tr}\, e^{-\beta(\hat{H} - \mu \hat{N})}, \tag{2.11}$$

where $\beta = 1/(k_B T)$ is the inverse temperature, and μ the chemical potential. $\hat{H}$ is the Hamiltonian of the system in the second quantized form,

$$\hat{H} = \sum_{\alpha} e_{\alpha} \hat{a}_{\alpha}^{\dagger} \hat{a}_{\alpha} + \sum_{\alpha, \beta, \gamma, \delta} \langle \alpha\beta | V | \gamma\delta \rangle \hat{a}_{\alpha}^{\dagger} \hat{a}_{\beta}^{\dagger} \hat{a}_{\delta} \hat{a}_{\gamma}, \tag{2.12}$$

and $\hat{N}$ the particle number operator,

$$\hat{N} = \sum_{\alpha} \hat{a}_{\alpha}^{\dagger} \hat{a}_{\alpha}. \tag{2.13}$$

Using Eq. (2.9) we first rewrite the partition function as

$$Z = \int \prod_{\alpha} \frac{d\Phi_{\alpha}^* d\Phi_{\alpha}}{2\pi i} e^{-\sum_{\alpha} \Phi_{\alpha}^* \Phi_{\alpha}} \langle \Phi | e^{-\beta(\hat{H} - \mu \hat{N})} | \Phi \rangle. \tag{2.14}$$

Dividing the imaginary time interval β into M pieces and inserting the unity operator in Eq. (2.14) $M - 1$ times, the partition function becomes

$$Z = \int \prod_{k=0}^{M-1} \prod_{\alpha} \frac{d\Phi_{\alpha,k}^* d\Phi_{\alpha,k}}{2\pi i} e^{-\sum_{k=0}^{M-1} \sum_{\alpha} \Phi_{\alpha,k}^* \Phi_{\alpha,k}} \prod_{k=1}^{M} \langle \Phi_{k-1} | e^{-\epsilon(\hat{H} - \mu \hat{N})} | \Phi_k \rangle, \tag{2.15}$$

where $\Phi_0 = \Phi_M = \Phi$, and $\epsilon = \beta/M$. For $\epsilon \ll 1$ the requisite matrix elements may be approximated as

$$\langle \Phi_{k-1} | e^{-\epsilon(\hat{H} - \mu \hat{N})} | \Phi_k \rangle = e^{-\epsilon \langle \Phi_{k-1} | \hat{H} - \mu \hat{N} | \Phi_k \rangle} + O(\epsilon^2). \tag{2.16}$$

Note that both the Hamiltonian and the number operator have all the creation operators to the left of all the annihilation operators. This is referred to as the "normal ordering". For an arbitrary normally ordered function of creation and annihilation operators A the definition of coherent states implies that

$$\langle \Phi | A(\hat{a}_{\alpha}^{\dagger}, \hat{a}_{\alpha}) | \Phi' \rangle = A(\Phi_{\alpha}^*, \Phi_{\alpha}') e^{\sum_{\alpha} \Phi_{\alpha}^* \Phi_{\alpha}'}. \tag{2.17}$$

Using Eqs. (2.16) and (2.17) and taking the limit $M \to \infty$, the partition function may be written as

$$Z = \lim_{M \to \infty} \int \prod_{k=0}^{M-1} \prod_{\alpha} \frac{d\Phi_{\alpha,k}^* d\Phi_{\alpha,k}}{2\pi i} e^{-\sum_{k=0}^{M-1} \sum_{\alpha} \Phi_{\alpha,k}^* (\Phi_{\alpha,k} - \Phi_{\alpha,k+1})}$$
$$\times e^{-\sum_{k=0}^{M-1} \epsilon(H(\Phi_{\alpha,k}^*, \Phi_{\alpha,k+1}) - \mu \sum_{\alpha} \Phi_{\alpha,k}^* \Phi_{\alpha,k+1})}. \tag{2.18}$$

If we interpret the index k as labeling moments in the "imaginary time", the last expression suggests that the partition function in the continuum limit may be understood as a *functional integral*:

$$Z = \int_{\Phi_\alpha(0)=\Phi_\alpha(\beta)} D\Phi^*_\alpha(\tau)D\Phi_\alpha(\tau)e^{-S[\Phi^*_\alpha(\tau),\Phi_\alpha(\tau)]}, \qquad (2.19)$$

where the *action* S adopts a simple form given by

$$S = \int_0^\beta d\tau \left[\sum_\alpha \Phi^*_\alpha(\tau)(-\partial_\tau - \mu)\Phi_\alpha(\tau) + H[\Phi^*_\alpha(\tau), \Phi_\alpha(\tau)] \right]. \qquad (2.20)$$

The measure of the functional integral in Eq. (2.19) should be understood as the "sum" over all complex functions $\Phi_\alpha(\tau)$ that satisfy the boundary condition $\Phi_\alpha(0) = \Phi_\alpha(\beta)$. The quantum number α can be anything that labels the states in the single-particle basis: momentum, position, lattice site in a system on a discrete lattice, etc. For example, choosing $\alpha = \vec{x}$ with $\vec{x}$ as the particle's continuous coordinate, the action for the system of bosons of mass m interacting via $V(\vec{x} - \vec{y})$ becomes

$$S = \int_0^\beta d\tau \int d\vec{x} \left[\Phi^*(\vec{x}, \tau) \left(-\partial_\tau - \mu - \frac{\hbar^2 \nabla^2}{2m} \right) \Phi(\vec{x}, \tau) \right.$$
$$\left. + \int d\vec{y} |\Phi(\vec{x}, \tau)|^2 V(\vec{x} - \vec{y})|\Phi(\vec{y}, \tau)|^2 \right]. \qquad (2.21)$$

The partition function for the system of interacting bosons may therefore be taken to be the sum over all possible complex functions of space and imaginary time which are *periodic* in imaginary time. The information about all the phase transitions in the system, and in particular about the superfluid transition, is in principle contained in the coherent-state functional integral in Eq. (2.19). As discussed in the next section, however, the partition function turns out not to be easily calculable, and one has to formulate a cunning strategy to address the physics of the superfluid phase transition.

2.2 Bose–Einstein condensation

Consider first the non-interacting system of bosons, with $V(\vec{r}) = 0$. Decomposing the periodic function as

$$\Phi(\vec{r}, \tau) = \frac{1}{\beta} \int \frac{d\vec{k}}{(2\pi)^d} \sum_{\omega_n} \Phi(\vec{k}, \omega_n)e^{i\vec{k}\cdot\vec{r}+i\omega_n\tau}, \qquad (2.22)$$

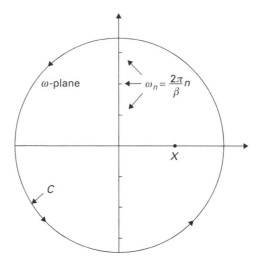

Figure 2.2 The integration contour used in the evaluation of the sum over Matsubara frequencies.

with the *Matsubara frequencies* $\omega_n = 2\pi n/\beta$, with n integer or zero, in terms of the Fourier components the path integral factorizes as

$$Z = Z_0 = \prod_{\vec{k},\omega_n} \int \frac{d\Phi^*(\vec{k},\omega_n)d\Phi(\vec{k},\omega_n)}{2\pi i} e^{-\beta^{-1}\sum_{\vec{k},\omega_n}(-i\omega_n+\frac{\hbar^2 k^2}{2m}-\mu)|\Phi(\vec{k},\omega_n)|^2}$$

$$= \prod_{\vec{k},\omega_n} \frac{\beta}{-i\omega_n + \frac{\hbar^2 k^2}{2m} - \mu}. \tag{2.23}$$

Let us see how this result leads to the familiar phenomenon of Bose–Einstein condensation. From the free energy $F_0 = -k_\mathrm{B}T \ln Z_0$ we may compute the number of particles as

$$N = -\frac{\partial F_0}{\partial \mu} = \frac{1}{\beta} \lim_{\eta \to 0} \sum_{\vec{k}} \sum_{\omega_n} \frac{e^{i\omega_n \eta}}{-i\omega_n + \frac{\hbar^2 k^2}{2m} - \mu}, \tag{2.24}$$

where we have included an infinitesimal η to render the frequency sum finite. The sum over Matsubara frequencies may be computed by using the calculus of residues as follows. Consider the integral

$$I = \int_C d\omega \frac{e^{\omega \eta}}{(e^{\omega \beta} - 1)(\omega - x)}, \tag{2.25}$$

where the contour C is a large circle in the complex ω-plane as in Fig. 2.2. In the infinite limit of the circle's radius the integrand, vanishes exponentially fast so that the integral $I = 0$. On the other hand, summing the contributions

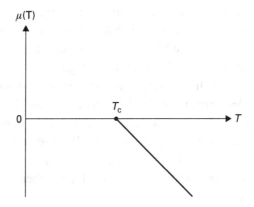

Figure 2.3 Temperature dependence of the chemical potential in a three-dimensional system of non-interacting bosons.

from the poles of the integrand, we find

$$I = 2\pi i \left(\frac{e^{x\eta}}{e^{\beta x} - 1} + \sum_{\omega_n = 2\pi n/\beta} \frac{e^{i\omega_n \eta}}{\beta(i\omega_n - x)} \right) \qquad (2.26)$$

for the infinitely large contour C. So,

$$\frac{1}{\beta} \sum_{\omega_n} \frac{e^{i\omega_n \eta}}{-i\omega_n + x} = \frac{e^{x\eta}}{e^{\beta x} - 1} = n_B(x), \qquad (2.27)$$

where $n_B(x)$ is the standard Bose occupation number, and we have set $\eta = 0$ in the last equation. Assuming that the occupation numbers vary smoothly with energy, in the thermodynamic limit we may also turn the remaining sum over wavevectors into an integral, and write

$$\frac{N}{V} = \int \frac{d\vec{k}}{(2\pi)^d} [e^{\frac{(\hbar^2 k^2/2m) - \mu}{k_B T}} - 1]^{-1}. \qquad (2.28)$$

For all the occupation numbers to be positive the chemical potential must be either negative or zero. At a fixed particle number $\mu(T)$ evidently also has to be a decreasing function. Assume that at some temperature T_{BEC}, $\mu(T_{BEC}) = 0$. The temperature dependence of the chemical potential will be then as in Fig. 2.3. In $d = 3$ and right at $T = T_{BEC}$, then

$$\frac{N}{V} = \int \frac{d\vec{k}}{(2\pi)^d} \frac{1}{e^{\frac{\hbar^2 k^2}{2m k_B T_{BEC}}} - 1} = 2.612 \left(\frac{2\pi m k_B T_{BEC}}{h^2} \right)^{3/2}. \qquad (2.29)$$

Since at all temperatures below T_{BEC} the chemical potential of the non-interacting system stays zero, the value of the integral for total number of particles in Eq. (2.28) is in fact at its finite maximum at $T = T_{BEC}$. If the number of particles is fixed, this means that at $T < T_{BEC}$ some of the single-particle states must be occupied by a finite fraction of the total number of particles to make up for the difference. This singular contribution has been missed by approximating the discrete sum over wavevectors with the integral, and needs to be added separately. As the occupation number is a decreasing function of energy, the macroscopically occupied state below T_{BEC} is the ground state at $\vec{k} = 0$. This is the phenomenon of Bose–Einstein condensation. The Bose–Einstein condensation temperature T_{BEC} is therefore determined by the density of particles and Eq. (2.29), and the number of particles in the condensate N_0 for $T < T_{BEC}$ is

$$\frac{N_0}{N} = 1 - \left(\frac{T}{T_{BEC}}\right)^{3/2}. \tag{2.30}$$

Note that in two dimensions the integral for the total number of particles is infinite for $\mu = 0$, and consequently there is no Bose–Einstein condensation at any finite temperature. Of course, at $T = 0$ all the bosons in the non-interacting system occupy the single-particle ground state. So we may say that in two dimensions $T_{BEC} = 0$. Bose–Einstein condensation in three dimensions is in fact the only phase transition that takes place in a non-interacting system.

Problem 2.1 Show that there is no Bose–Einstein condensation in the system with a discrete energy spectrum.

Solution If the spectrum is discrete, assuming $\mu = e_0$ at $T \neq 0$, with e_0 as the single-particle ground state, yields an infinite total number of particles. So just as in the case of free bosons in two dimensions, there is no Bose–Einstein condensation at any finite temperature, and $\mu(T) < e_0$. Only when there is a continuum of low-energy excitations, as in the thermodynamic limit of particles in a box, can the chemical potential reach the ground state energy at a finite temperature and still yield an integrable singularity for the particle density. For the same reason there is no Bose–Einstein condensation in finite-size systems.

2.3 Hartree approximation

Let us include now the interaction term, and approximate the two-body inter-action as $V(\vec{r}) \to \lambda \delta(\vec{r})$. This simplification will prove to be completely sufficient for the understanding of the critical behavior near the superfluid transition. In Fourier space, the approximation corresponds to replacing the

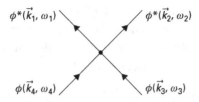

Figure 2.4 Diagrammatic representation of the interaction term in Eq. (2.31). The arrows pointing inwards represent Φs, pointing outwards Φ^*s, and the sum of incoming wavevectors (frequencies) equals the sum of outgoing wavevectors (frequencies).

interaction $V(\vec{k})$ by the constant λ. Since this is clearly inadequate for large wavevectors where $V(k)$ has to approach zero, we will restrict all the integrals over wavevectors to $|\vec{k}| < \Lambda$, where Λ is an arbitrary cutoff. Its precise value will turn out to be unimportant for the critical behavior, and we may assume that it is of the order of inverse separation between the particles, which provides the natural short length scale in the problem. This means that our approximation is adequate only at distances long compared to the separation between atoms. Such an "ultraviolet" cutoff will be then implicitly assumed in all the calculations hereafter.

It is evident why the partition function is not so easily calculable once $\lambda > 0$. In terms of the Fourier components the quartic term in the action becomes

$$\lambda \int_0^\beta d\tau \int d\vec{r} |\Phi(\vec{r}, \tau)|^4$$

$$= \frac{\lambda}{\beta^4} \int \frac{d\vec{k}_1 \cdots d\vec{k}_4}{(2\pi)^{3d}} \sum_{\omega_1, \ldots, \omega_4} \delta_{\omega_1 + \omega_2, \omega_3 + \omega_4} \delta(\vec{k}_1 + \vec{k}_2 - \vec{k}_3 - \vec{k}_4)$$

$$\times \Phi^*(\vec{k}_1, \omega_1) \Phi^*(\vec{k}_2, \omega_2) \Phi(\vec{k}_3, \omega_3) \Phi(\vec{k}_4, \omega_4), \tag{2.31}$$

so that the functional integral no longer factorizes into a product over simpler integrals as in Eq. (2.23). If we remained in the original space-time representation in Eq. (2.21), on the other hand, the quartic term would be "diagonal", but the quadratic term would couple different points in space and imaginary time, and the factorization would again be impossible. With the interaction term present one in general cannot find the quantum numbers which would factorize the path integral. This is a consequence of the quantum mechanical uncertainty relation between the position and the momentum operators. This is exactly why the many-body problems, exemplified by the action in Eq. (2.21), are generally non-trivial. The quartic term may be represented by the diagram in Fig. 2.4.

One may try to estimate the effect of interactions on the transition by formulating a self-consistent mean-field theory, similar to Curie–Weiss theory of the Ising model in Problem 1.4. Instead of dealing with the quartic term in Eq. (2.31) we replace it by an effective quadratic term as

$$|\Phi(\vec{r}, \tau)|^4 \rightarrow \langle|\Phi(\vec{r}, \tau)|^2\rangle|\Phi(\vec{r}, \tau)|^2, \tag{2.32}$$

where the average $\langle|\Phi(\vec{r}, \tau)|^2\rangle$ is to be self-consistently computed. This is the so-called "Hartree approximation". The effect of interaction in this approximation is thus only to shift the chemical potential μ to $\tilde{\mu}$:

$$\mu \rightarrow \tilde{\mu} = \mu - \lambda\langle|\Phi(\vec{r}, \tau)|^2\rangle \tag{2.33}$$

in the partition function for the *effective* non-interacting system, as in Eq. (2.23). Computing the above average yields then an equation for the effective chemical potential:

$$\tilde{\mu} = \mu - \frac{\lambda}{\beta}\sum_{\omega_n}\int\frac{d\vec{k}}{(2\pi)^d}\frac{e^{i\omega_n\eta}}{-i\omega_n + \frac{k^2}{2m} - \tilde{\mu}}, \tag{2.34}$$

where we introduced again an infinitesimal η.

The Bose–Einstein condensation now occurs not at $\mu = 0$ but at $\tilde{\mu} = 0$, which corresponds to

$$\mu = \mu_c(T) = \lambda\int\frac{d\vec{k}}{(2\pi)^d}\left[e^{\frac{\hbar^2 k^2}{2mk_B T}} - 1\right]^{-1}. \tag{2.35}$$

For $d > 2$ the integral over the wavevectors is finite. Rescaling the wavevectors with temperature, we see that at low temperatures the transition line behaves like $\mu_c(T) \propto T^{\frac{d}{2}}$. Taking the chemical potential μ to be the tuning parameter, at $T = 0$ the condensation occurs at $\mu_c(0) = 0$. We will refer to such transitions at $T = 0$ as *quantum phase transitions*. Quantum phase transitions are tuned by changing a coupling constant in the Hamiltonian, like the chemical potential or the interaction, and correspond to a non-analytic change of the ground state energy. They may also be discontinuous or continuous, the latter being the case in our example.

At $T = 0$ Matsubara frequencies form a continuum. At finite temperature, on the other hand, the action for the Fourier modes with $\omega_0 = 0$ is separated by a finite amount $\sim T$ from the action for the rest of the modes with $\omega_n \neq 0$. Imagine performing the integral over all the modes in the partition function except those with ω_0 at a finite temperature. Since all such modes have the quadratic coefficient in the action $2\pi T$ or larger, such integration may provide only finite modifications of the coefficients in the remaining action for the ω_0

modes, besides yielding an analytic contribution to the free energy. As will be discussed further in the next chapter, the actual values of the coefficients in the action are typically unimportant for the critical behavior, and determine only the non-universal quantities like the transition temperature. ω_n modes with $|n| \geq 1$ at $T > 0$ may therefore be named *non-critical*, as they do not affect the universal aspects of the critical behavior at the finite temperature transition (see Problem 2.2).

Another way to see this is to recall that all the singularities of thermodynamic or correlation functions come from the integration over small wavevectors, i.e. from the thermodynamic limit. At finite temperature the partition function in Eq. (2.19) may be thought of as describing a system of finite extent in the direction of imaginary time, and infinite in spatial directions. Compared to long length scales (corresponding to small wavevectors) that matter for the critical behavior, the finite size along the imaginary time direction is negligible, and the complex field may be taken to be uniform along that direction. This is tantamount to neglecting all but the modes with $\omega_0 = 0$ Matsubara frequency.

At $T = 0$ the gap between the zero and the finite frequency modes collapses and all the modes need to be taken into account at the quantum critical point. Quantum critical behavior therefore often resembles the finite temperature behavior, but in an effectively *higher* dimension. This issue will be discussed in detail in Chapter 8.

Finally, note that for dimensions $d \leq 2$ the integral in Eq. (2.35) diverges at small wavevectors. The self-consistent approximation would suggest that there is no finite temperature transition in $d \leq 2$, similarly to the absence of the Bose–Einstein condensation in the non-interacting system. This will indeed prove to be correct, except for $d = 2$ when there is a particular Kosterlitz–Thouless transition. This however, lies outside the reach of the Hartree approximation, and will be the subject of Chapter 6.

Problem 2.2 Calculate the equal-time correlation function

$$\chi(\vec{x}) = \langle \Phi(0, \tau) \Phi^*(\vec{x}, \tau) \rangle$$

in the system of non-interacting bosons, at large $|\vec{x}|$.

Solution In terms of Fourier components the correlation function becomes

$$\chi(\vec{x}) = \frac{1}{\beta^2} \sum_{\omega_n, \omega_n'} \int \frac{d\vec{k}\, d\vec{k}'}{(2\pi)^{2d}} e^{i(\omega_n - \omega_n')\tau - i\vec{k}'\cdot\vec{x}} \langle \Phi(\vec{k}, \omega_n) \Phi^*(\vec{k}', \omega_n') \rangle,$$

with

$$\langle \Phi(\vec{k}, \omega_n)\Phi^*(\vec{k}', \omega_n') \rangle = \frac{\beta(2\pi)^d}{-i\omega_n + \frac{\hbar^2 k^2}{2m} - \mu} \delta_{\omega_n, \omega_n'} \delta(\vec{k} - \vec{k}')$$

in the non-interacting system. Performing the sum over the Matsubara frequencies yields

$$\chi(\vec{x}) = \int \frac{d\vec{k}}{(2\pi)^d} \frac{e^{-i\vec{k}\cdot\vec{x}}}{e^{\frac{(\hbar^2 k^2/2m) - \mu}{k_B T}} - 1} + n_0(T)\theta(T_{\text{BEC}} - T),$$

where the condensate contribution $n_0(T) = N_0(T)/V$ as in Eq. (2.30) must be added separately at $T < T_{\text{BEC}}$.

After performing the angular integrals in $d = 3$, using the calculus of residues the correlation function at $T > T_{\text{BEC}}$ may be written as

$$\chi(\vec{x}) = \frac{mk_B T}{2\pi \hbar^2 |\vec{x}|} \text{Re} \sum_{k_n} e^{ik_n |\vec{x}|},$$

where $\{k_n\}$ determined by $(\hbar^2 k_n^2/2m) + |\mu| = i\omega_n$, with $n = 0, \pm 1, \pm 2, \ldots$, are the simple poles of the integrand in the complex k-plane, and the sum is to be performed only over the poles with positive imaginary parts. For large $|\vec{x}|$ the dominant contribution evidently comes from the pole closest to the real axis, i.e. for $n = 0$. In this limit, therefore,

$$\chi(\vec{x}) \approx \frac{mk_B T}{2\pi \hbar^2} \frac{e^{-\frac{|\vec{x}|\sqrt{2m|\mu|}}{\hbar}}}{|\vec{x}|} + n_0(T)\theta(T_{\text{BEC}} - T).$$

At temperatures $T > T_{\text{BEC}}$ correlations decay exponentially quickly at large separations, whereas for $T < T_{\text{BEC}}$ there is a true long range order, and

$$\lim_{|\vec{x}|\to\infty} \chi(\vec{x}) = n_0(T).$$

Note also that the result at $T > T_{\text{BEC}}$ could have been obtained by neglecting all but $\omega_n = 0$ contribution to the sum over Matsubara frequencies.

2.4 Landau's mean-field theory

Let us consider the finite temperature transition. In accordance with the previous discussion we omit the regular (analytic) piece of the free energy coming from the modes with $\omega_n > 0$, and retain only the $\omega_0 = 0$ modes in the partition function. The fields are then independent of the imaginary time and the partition function becomes $Z = \int D\Phi^*(\vec{r})D\Phi(\vec{r})e^{-S}$ with the action

$$S[\Phi] = \frac{1}{k_B T} \int d\vec{r} \left[\frac{\hbar^2}{2m} |\nabla\Phi(\vec{r})|^2 - \mu|\Phi(\vec{r})|^2 + \lambda|\Phi(\vec{r})|^4 \right]. \quad (2.36)$$

This partition function defines the *Ginzburg–Landau–Wilson theory*. For the superfluid transition the fluctuating field $\Phi(\vec{r})$ is a complex number, i.e. it has *two* real components. One can consider a more general field with N real components, with the action of the same form as in Eq. (2.36). For $N = 1$ such an action would be symmetric only with respect to an overall change of sign of the fields, $\Phi(\vec{r}) \to -\Phi(\vec{r})$. This is the same global Z_2 symmetry as present in the Ising model, and we expect that the two models will lie in the same universality class. The action in Eq. (2.36) with $N = 1$ real field should describe then the ferromagnetic transition in the Ising model. For $N = 3$ real components the action is symmetric under an arbitrary global rotation of the fields, $\Phi_i(\vec{r}) \to M_{ij}\Phi_j(\vec{r})$, with the real matrix $M^T M = 1$, and thus belonging to the group $SO(3)$. Ginzburg–Landau–Wilson theory with three real components thus has the same symmetry as the Heisenberg model and thus is expected to describe the magnetic system of Heisenberg spins near the critical point. Besides the superfluid transition, $N = 2$ theory also describes the magnetic system in which magnetization is confined to a plane, i.e. the XY model defined in Section 1.2. Most of our considerations will directly generalize to a general number of components, and will be equally relevant to magnetic and the superfluid phase transitions.

The Ginzburg–Landau–Wilson partition function with general number of components therefore serves as a sort of a meta-model, that is a general model to which more specific models like the Ising, the XY, and the Heisenberg model reduce in the critical region. Only the universal critical behavior, and not the thermodynamics outside the critical region, may be expected to be given correctly by the Ginzburg–Landau–Wilson theory. The reason for this, which is essentially the explanation of universality, will become clear after Chapter 3.

As a first step towards the systematic theory of critical phenomena we may approximate the partition function in the Ginzburg–Landau–Wilson theory with the value of the integrand at its maximum. In such a *saddle-point approximation* to the path integral the free energy of the system is simply $F = k_B T S[\Phi_0]$, where Φ_0 is the configuration that minimizes the action, determined by

$$\frac{\delta S}{\delta \Phi}\Big|_{\Phi_0} = 0. \quad (2.37)$$

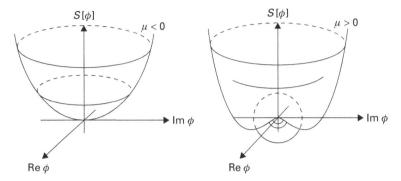

Figure 2.5 The action S as a function of the constant complex field Φ. For $\mu < 0$ there is a unique minimum at $\Phi = 0$, whereas at $\mu > 0$ S resembles a "Mexican hat", and develops a continuum of minima at finite Φ.

Obviously Φ_0 will be independent of the coordinate, and be either zero or

$$|\Phi_0|^2 = \frac{\mu}{2\lambda}. \tag{2.38}$$

Taking the chemical potential to be the tuning parameter, the system is in the normal phase for $\mu < 0$, when the non-trivial solution in Eq. (2.38) is impossible, so $\Phi_0 = 0$. When $\mu > 0$, on the other hand, Φ_0 is finite, since

$$S[\Phi_0] = -\frac{\mu^2}{4\lambda} \frac{V}{k_\mathrm{B} T} < S[0] = 0. \tag{2.39}$$

In terms of the temperature, the transition in the saddle-point approximation occurs at the temperature T_c where the function $\mu(T)$ changes sign. We may therefore understand $-\mu$ to be a general coefficient of the quadratic term in the action, which changes sign near the transition: $\mu \sim (T_\mathrm{c} - T) + O((T - T_\mathrm{c})^2)$.

For $\mu > 0$, $\Phi_0 = 0$ becomes a local maximum of the action, as depicted in Fig. 2.5. Since the finite value of Φ_0 signals the ordered phase, it can be taken to be the order parameter for the superfluid phase transition. In general, the order parameter is defined as the average $\langle \Phi(\vec{r}) \rangle$, which in the saddle-point approximation coincides with Φ_0. Note that only the absolute value of Φ_0 is fixed by the condition in Eq. (2.38), whereas its phase is left arbitrary. For convenience, it may be chosen to be zero. This, or any other choice, breaks the global $U(1)$ invariance under the transformation

$$\Phi(\vec{r}) \to \mathrm{e}^{\mathrm{i}\phi} \Phi(\vec{r}) \tag{2.40}$$

present in the action (2.36). We may say that the $U(1)$ symmetry is *spontaneously broken* in the ordered (superfluid) phase, and preserved in the disordered (normal) phase. The $U(1)$ symmetry here is an analog of the familiar rotational symmetry that becomes spontaneously broken by the direction of the finite magnetization in the example of ferromagnetic phase transition. Consequences of spontaneous breaking of a continuous symmetry will be explored in detail in Chapter 5.

Since $\mu \sim T_c - T$ in the Ginzburg–Landau–Wilson theory, differentiating $S[\Phi_0]$ twice with respect to temperature, the specific heat is found to have only a finite discontinuity at the transition, and the critical exponent $\alpha = 0$. To compute other critical exponents in the saddle-point approximation we first define the superfluid susceptibility as

$$\chi(\vec{r} - \vec{r}') = \frac{\delta^2 \ln Z[j]}{\delta j^*(\vec{r})\delta j(\vec{r}')}|_{j \equiv 0},\qquad(2.41)$$

with the partition function $Z[j]$ defined as before but with the action in Eq. (2.36) changed as

$$S \to S + \int d\vec{r}[\Phi(\vec{r})j(\vec{r}) + \Phi^*(\vec{r})j^*(\vec{r})].\qquad(2.42)$$

Performing the differentiations, the susceptibility becomes

$$\chi(\vec{r} - \vec{r}') = \langle \Phi^*(\vec{r})\Phi(\vec{r}')\rangle - |\langle\Phi(\vec{r})\rangle|^2,\qquad(2.43)$$

where we used the fact that the system is translationally invariant, so that χ is a function only of the difference in coordinates. The source field $j(\vec{r})$ is introduced to couple directly to Φ in the action in the same way as the real magnetic field would enter the partition function of magnetic systems. For the superfluid transition, however, the source field is just a mathematical device for generating the response function without a direct physical meaning.

The susceptibility may be easily computed in the saddle-point approximation by considering small fluctuations of $\Phi(\vec{r})$ around the saddle point. Assume $\mu < 0$ and expand the action around the trivial saddle point, in the normal phase. The result is again Eq. (2.36). In the first approximation we may set $\lambda = 0$, so that the action becomes quadratic in the fluctuating fields. $Z[j]$ can then be straightforwardly computed by simply completing the square in the exponent. The result is:

$$\ln Z[j] = \int d\vec{r}\, d\vec{r}'\, j^*(\vec{r})\chi_0(\vec{r} - \vec{r}')j(\vec{r}') + \text{constant},\qquad(2.44)$$

where the constant is independent of the source field. The mean-field susceptibility in the normal phase, which equals the susceptibility for the non-interacting system, is then

$$\chi_0(\vec{r}) = k_B T \int \frac{d\vec{k}}{(2\pi)^d} \frac{e^{i\vec{k}\cdot\vec{r}}}{\frac{\hbar^2 k^2}{2m} - \mu}. \tag{2.45}$$

Near the critical temperature T_c at which $\mu(T_c) = 0$ we can rewrite this as

$$\chi_0(\vec{r}) = \frac{2m k_B T_c}{\hbar^2} \frac{F(r/\xi)}{r^{d-2}}, \tag{2.46}$$

with the scaling function

$$F(z) = z^{d-2} \int \frac{d\vec{q}}{(2\pi)^d} \frac{e^{i\vec{q}\cdot\vec{z}}}{q^2 + 1}, \tag{2.47}$$

and the correlation length $\xi = \hbar/\sqrt{2m|\mu|}$. Note that the result for the suscep-
tibility may indeed be cast in the scaling form in Eq. (1.26). Since $\mu \propto T_c - T$
near T_c, the correlation length exponent is $\nu = 1/2$, and the anomalous dimen-
sion $\eta = 0$. In the Fourier space $\chi(\vec{k} = 0) \sim |\mu|^{-1}$, and thus the exponent
$\gamma = 1$.

The saddle-point approximation to our functional integral therefore leads
to the already familiar mean-field values of the critical exponents. In fact, the
approximation we employed is nothing else but the general mean-field theory
of phase transitions introduced first phenomenologically by L. Landau. The
critical exponents are the same as in the Curie–Weiss theory of magnetism and
in the van der Waals–Maxwell theory of the liquid–gas transition. In particular,
they are completely independent of dimensionality d and the number of field
components N.

Problem 2.3* Find the critical exponents in the model with

$$Z = \int \prod_{i=1}^{N} D\Phi_i^*(\vec{r}) D\Phi_i(\vec{r}) e^{-S}$$

with

$$S = \int d\vec{r} \left[\sum_{i=1}^{N} \left(|\nabla\Phi_i(\vec{r})|^2 + t|\Phi_i(\vec{r})|^2 \right) + \frac{\lambda}{2N} \left(\sum_{i=1}^{N} |\Phi_i(\vec{r})|^2 \right)^2 \right],$$

and the ultraviolet cutoff Λ, in general dimension and when $N \to \infty$.

Solution This is an example where the Hartree approximation discussed in
the text yields an exact solution. The interaction term can be decoupled using

the *Hubbard–Stratonovich transformation* (Appendix A)

$$e^{-\frac{\lambda}{2N}\int d\vec{r}|\Phi(\vec{r})|^4} = \int D\chi(\vec{r}) e^{-\int d\vec{r}[\frac{N}{2\lambda}\chi^2(\vec{r})+i\chi(\vec{r})|\Phi(\vec{r})|^2]},$$

where $|\Phi(\vec{r})|^2 = \sum_{i=1}^N |\Phi_i(\vec{r})|^2$. After this transformation the action becomes quadratic in Φ_i, which can therefore in principle be integrated out. Anticipating the condensation transition, we integrate out all the components Φ_i, $2 \le i \le N$, except the first. Since N can be pulled out in front of the action in the exponent, in the limit $N \to \infty$ the saddle-point approximation to the partition function becomes exact. Assuming the fields at the saddle-point to be uniform, after rotating $i\chi \to \chi$, and introducing $\sigma = \Psi_1/N^{1/2}$, the free energy per unit volume becomes

$$\frac{F}{Nk_\mathrm{B}T} = (t + \chi)|\sigma|^2 - h\sigma^* - h^*\sigma - \frac{\chi^2}{2\lambda} + \int \frac{d\vec{k}}{(2\pi)^d} \ln(k^2 + t + \chi),$$

with

$$(t + \chi)\sigma = h,$$

and

$$\chi = \lambda \int \frac{d\vec{k}}{(2\pi)^d} \frac{1}{k^2 + t + \chi} + \lambda|\sigma|^2.$$

We have also introduced an external field h to couple to the order parameter.

Let us first set $h = 0$. In the normal phase $\sigma = 0$, and thus $t + \chi > 0$. The transition temperature is given by

$$t_\mathrm{c} = -\lambda \int \frac{d\vec{k}}{(2\pi)^d} \frac{1}{k^2},$$

which is finite for $d > 2$. For $d \le 2$ there is no transition. For $d > 2$ we may write

$$(t + \chi)\left(1 + \lambda \int \frac{d\vec{k}}{(2\pi)^d} \frac{1}{k^2(k^2 + t + \chi)}\right) = t - t_\mathrm{c}.$$

With the ultraviolet cutoff Λ in the integral over wavevectors, as $t + \chi \to 0$ the last integral is finite for $d > 4$ and we find $t + \chi \propto (t - t_\mathrm{c})$. Since $k^2 + t + \chi$ is just the inverse susceptibility, the exponent $\gamma = 1$ for $d > 4$. For $2 < d < 4$, on the other hand, the integral diverges like $\sim (t + \chi)^{(d-4)/2}$. Therefore $\gamma = 2/(d - 2)$ for $2 < d < 4$. Exactly in $d = 4$ the integral is logarithmically divergent and $(t + \chi) \sim (t - t_\mathrm{c})/|\ln(t - t_\mathrm{c})|$. Since anomalous dimension $\eta = 0$ in all dimensions, the correlation length exponent $\nu = \gamma/2$.

When $\sigma \neq 0$ and $h = 0$, $\chi = -t$. The uniform susceptibility is therefore infinite in the condensed phase. Then

$$|\sigma|^2 = \frac{t_c - t}{\lambda},$$

and the exponent $\beta = 1/2$ in all dimensions.

Finally, at $h > 0$ and $t = t_c$ one finds

$$(t_c + \chi)^3 \left(1 + \lambda \int \frac{d\vec{k}}{(2\pi)^d} \frac{1}{k^2(k^2 + t_c + \chi)}\right) = \lambda h^2,$$

and thus $(t_c + \chi) \propto h^{2/3}$ for $d > 4$, and $(t_c + \chi) \propto h^{4/(2+d)}$ for $2 < d < 4$. Since $\sigma = h/(t_c + \chi)$, the exponent $\delta = 3$ for $d > 4$, and $\delta = (d + 2)/(d - 2)$ for $2 < d < 4$. Differentiating the free energy twice with respect to temperature in the normal phase one finds a discontinuity ($\alpha = 0$) for $d > 4$, and $\alpha = (d - 4)/(d - 2)$ for $2 < d < 4$. The hyperscaling is satisfied for $2 < d < 4$, and violated above four dimensions. In the condensed phase, however, the specific heat is always finite.

Problem 2.4* Consider the quantum phase transition at $T = 0$ in the theory with the action

$$S = \int_0^\beta d\tau \int d\vec{r} \left[|\partial_\tau \Phi(\vec{r}, \tau)|^2 + |\nabla \Phi(\vec{r}, \tau)|^2 - \mu |\Phi(\vec{r}, \tau)|^2 \right.$$
$$\left. + \frac{\lambda}{N} |\Phi(\vec{r}, \tau)|^4 \right],$$

where Φ is the N-component complex field, like in the previous problem. Find the critical exponents in general dimension when $N \to \infty$.

Solution The exponents are the same as in Problem 2.3 but with d replaced by $d + 1$. There is no transition for $d \leq 1$, and the critical exponents have the mean-field values for $d > 3$.

2.5 Upper critical dimension

Landau's mean-field theory, at least in the normal phase, may be understood as the zeroth order approximation in the interaction strength λ. It seems plausible then that successively better approximations for susceptibility, for example, would be obtained by performing a perturbation theory in λ. This indeed is true everywhere except in the critical region, where a direct perturbative approach is doomed to failure. This may be seen by finding the dimensionless parameter in terms of which such a perturbation theory would be formulated. We may express the Ginzburg–Landau–Wilson action in Eq. (2.36) in terms of the dimensionless fields, lengths, and interaction, defined as $\Psi = \xi^{(d-2)/2}\Phi/\Lambda$, $\vec{z} = \vec{r}/\xi$, and $\hat{\lambda} = (2m\Lambda^2/\hbar^2)\lambda\xi^{4-d}$, with the correlation length $\xi = \hbar/\sqrt{2m|\mu|}$. The action for $\mu < 0$ then becomes

$$S = \frac{\hbar^2\Lambda^2}{2mk_BT} \int d\vec{z} \left[\left| \frac{\partial \Psi}{\partial \vec{z}} \right|^2 + |\Psi|^2 + \hat{\lambda}|\Psi|^4 \right]. \tag{2.48}$$

It is then evident that for the perturbation theory to succeed it is the dimensionless parameter $\hat{\lambda}$ that needs to be small. Near the critical point where the correlation length diverges, this, however, critically depends on the dimensionality of the system. So we need to distinguish two different cases.

(1) For $d > 4$, as $|\mu| \rightarrow 0, \hat{\lambda} \rightarrow 0$, and the mean-field theory yields the correct critical behavior.
(2) For $d < 4$, on the other hand, as $|\mu| \rightarrow 0, \hat{\lambda} \rightarrow \infty$, and perturbation theory necessarily breaks down in the critical region. It is still sensible away from the critical point, and this is the foundation of the high-temperature expansion. Near T_c, on the other hand, there is no obvious small parameter in the problem.

Four dimensions therefore separate the regimes where the critical behavior is determined exactly by the mean-field theory from where it is non-trivial. Such a special dimensionality is called the *upper critical dimension*, and for the Ginzburg–Landau–Wilson action in Eq. (2.36) it is $d_u = 4$. Note that the upper critical dimension does not depend on the number of components of the field Φ, and thus $d_u = 4$ also for the magnetic systems in the Ising and the Heisenberg universality classes. It does depend on some other features of the action (see Problem 2.5), and for some phase transitions it may be different from four.

Precisely at $d = 4$ the dimensionless coupling $\hat{\lambda}$ becomes independent of the correlation length, and so neither vanishes nor diverges in the critical region. We will see in the next chapter that in this special case the mean-field is still almost correct, up to some logarithmic, and therefore rather weak, corrections to the usual scaling.

Problem 2.5 Determine the upper critical dimension for the quantum $(T = 0)$ phase transitions described by the action

$$S = \int_0^\beta d\tau \int d\vec{x} \left[\left[\int d\vec{y} \int_0^\beta d\tau' \Phi(\vec{x}, \tau) K(\vec{x} - \vec{y}, \tau - \tau') \Phi(\vec{y}, \tau') \right] \right.$$
$$\left. + a\Phi^2(\vec{x}, \tau) + b\Phi^4(\vec{x}, \tau) \right],$$

where Φ has N real components, $\Phi^2 = \sum_{i=1}^N \Phi_i^2$, and the Fourier transform of the kernel $K(x, \tau)$ for small frequency and the wavevector is given by:

(a) $K(\omega, k) = c|\omega| + dk^2$ (describes the $T = 0$ antiferromagnetic transition for itinerant electrons, when $N = 3$),
(b) $K(\omega, k) = ic\omega + dk^2$ (superfluid transition when $N = 2$),
(c) $K(\omega, k) = c(|\omega|/k) + dk^2$ (ferromagnetic transition for itinerant electrons when $N = 3$),
(d) $K(\omega, k) = c(|\omega|/k^2) + dk^2$ (ferromagnetic transition for dirty itinerant electrons for $N = 3$),
(e) $K(\omega, k) = c\omega^2 + dk^2$ (superfluid transition for bosons on a lattice and at a commensurate density when $N = 2$), and
(f) $K(\omega, k) = c|\omega| + d\omega^2 + ek^2$ (superconducting transition for dirty d-wave superconductor when $N = 2$),

where c, d, and e are constants.

Solution If the the leading frequency and the momentum terms in the kernel K scale as $\omega \sim k^z$, the upper critical dimension for the quantum phase transition is $d_u = 4 - z$. The answer is therefore (a) 2, (b) 2, (c) 1, (d) 0, (e) 3, (f) 2.

3

Renormalization group

Wilson's momentum-shell transformation with the concomitant expansion around the upper critical dimension is defined. The basic notions of relevant and irrelevant couplings, renormalization flow, and fixed points are introduced. The origins of scaling and of universality are explained, and corrections to mean-field values of critical exponents are computed. The field theoretic renormalization group is briefly discussed and used to calculate the anomalous dimension.

3.1 Idea

We found that the direct perturbation theory in the Ginzburg–Landau–Wilson theory breaks down below the upper critical dimension because the perturbation parameter grows with the correlation length, and so becomes arbitrarily large as the critical point is approached. If the system were finite, on the other hand, the correlation length would be bound by its size, and perturbation theory could succeed. Singular thermodynamic behavior near the critical point comes from the thermodynamic limit, or, more precisely, from those modes that have arbitrary low energies in an infinitely large system. This is called the *infrared singularity*. This observation suggests the following strategy to avoid the problem of direct perturbation theory.

First, note that mass m only provides the energy scale in Eq. (2.36). It is practical to rescale it out by absorbing it into the chemical potential, redefined as $2m\mu/\hbar^2 \to \mu$, and into the interaction coupling, as $2m\lambda/\hbar^2 \to \lambda$. Similarly, near the critical point, the temperature $T \approx T_c$ may be replaced by the critical temperature, and then eliminated by rescaling the action as $2mk_B T_c S/\hbar^2 \to S$. After these simplifications the superfluid susceptibility becomes a function only of the wavevector, the (rescaled) chemical potential, and the interaction:

$$\chi(\vec{k}) = F(k, \mu, \lambda). \tag{3.1}$$

Implicit in this expression is still the dependence on the ultraviolet cutoff Λ, the inverse of which defines the shortest length scale in the problem. Since it is the "slow" modes with low wavevectors that lead to the singular critical behavior, we leave those and imagine integrating only over the "fast" modes with large wavevectors, $\Lambda/b < k < \Lambda$ with $b \approx 1$. Assume that after this integration the partition function for the remaining slow modes with $k < \Lambda/b$ can be recast in the same form as before the integration, except that the values of μ and λ became changed as

$$\mu \rightarrow \mu(b), \tag{3.2}$$
$$\lambda \rightarrow \lambda(b). \tag{3.3}$$

Under this, which must seem a rather strong assumption, the susceptibility for low wavevectors satisfies

$$\chi(\vec{k}) = b^x F(bk, \mu(b), \lambda(b)). \tag{3.4}$$

Multiplication of the wavevector by the factor b is due to the change of the shortest length scale in the problem caused by the integration over the fast modes, when the ultraviolet cutoff Λ is replaced by Λ/b. The factor b^x reflects the fact that in units of length χ has the dimension x, i.e. $\chi \sim \Lambda^{-x}$. The question now becomes what would happen as $b \rightarrow \infty$ and more and more modes are integrated this way. If

$$\lim_{b \to \infty} \lambda(b) = \lambda^*, \tag{3.5}$$

then for large b,

$$\chi(\vec{k}) = b^x F(bk, \mu(b), \lambda^*). \tag{3.6}$$

If the value of λ^* happens to be small we may now use perturbation theory in λ^*. The idea is therefore to avoid writing the perturbation series in the original interaction $\hat{\lambda}$ that diverges near the critical point, and instead determine the effective *renormalized* coupling for the low-energy modes first; see if that is small, and if it is use that renormalized coupling as the new perturbation parameter.

The success of this strategy is by no means guaranteed, and several things may go wrong.

(1) As b increases there could be new terms appearing in the action for the slow modes which were not present initially, such as, for example:

$$\lambda'(b)|\Phi|^6, \lambda''(b)|\Phi|^8, \lambda'''(b)|\nabla\Phi|^2|\Phi|^2. \tag{3.7}$$

All the generated terms must be still invariant under the symmetries of the original action. This prohibits the appearance of the terms such as $\Phi^*\Phi^2$, for example, which would violate the global $U(1)$ symmetry of the action. The same observation applies to other symmetries and limits the number of terms that can be generated by the integration over the fast modes.

As $b \to \infty$ such a new coupling *generated* by the integration over the high energy modes may approach zero or a finite value. If the limit is zero the coupling will be called *irrelevant*, and we were correct in neglecting it in the first place. If the limit is finite it needs to be included from the beginning and its evolution with parameter b followed. Such a coupling will be called *relevant*. The theory which has a finite number of such relevant couplings is called *renormalizable*. If the theory is *non-renormalizable* the above strategy will most likely fail. Whether a coupling is relevant or irrelevant will depend on the value of λ^* if more than one exists in the theory, as typically is the case.

(2) The value of λ^* may turn out to be finite but not small, so perturbation theory in λ^* seems not feasible. This indeed turns out to be the case for many physical situations, including the superfluid and the magnetic transitions. Nevertheless, this problem can be dealt with and the critical behavior determined quite accurately even when λ^* is of order unity. This is done by computing the series in λ^* to as high order as possible, and then making an educated guess for the function that is being expanded from the finite number of computed terms.

(3) There may not be a finite λ^* in the theory, but $\lambda(b)$ goes to infinity with increasing b. This sometimes indicates a first-order transition. An important physical example is provided by the type-I superconductors, described in the next chapter.

With these possible pitfalls in mind, we will proceed by assuming that λ^* exists and is finite. Furthermore, for $b \gg 1$ and for small $\mu < 0$ we will assume $\mu(b) \approx \mu b^y$. Choosing then the parameter b so that $\mu b^y = \mu_0$ with μ_0 constant, we may write the susceptibility for large b as

$$\chi(\vec{k}) = \left|\frac{\mu_0}{\mu}\right|^{x/y} F\left(k\left|\frac{\mu_0}{\mu}\right|^{1/y}, \mu_0, \lambda^*\right). \tag{3.8}$$

After the Fourier transform is taken this implies $\chi(\vec{r})$ in the scaling form introduced in Chapter 1. In particular, the correlation length and the susceptibility exponents are simply

$$\nu = \frac{1}{y},\tag{3.9}$$

$$\gamma = \frac{x}{y}.\tag{3.10}$$

Scaling laws then imply the values of the remaining four critical exponents. What is needed therefore are the functions $\lambda(b)$ and $\mu(b)$ which would allow us to compute the numbers x and y. These numbers, in principle, will be dependent on the value of λ^*.

3.2 Momentum-shell transformation

Let us implement the above strategy at the finite-temperature partition function in Eq. (2.36). We will assume that μ and λ have been rescaled so that one may set $\hbar^2/2m = k_B T_c = 1$, and consider the fields independent of the imaginary time containing the Fourier components with $|\vec{k}| < \Lambda$ only.

The field $\Phi(\vec{r})$ may be divided into slow and fast parts as

$$\Phi(\vec{r}) = \Phi_<(\vec{r}) + \Phi_>(\vec{r}),\tag{3.11}$$

where $\Phi_<$ contains only the Fourier components with $k < \Lambda/b$, and $\Phi_>$ only with $\Lambda/b < k < \Lambda$. The partition function may then be rewritten as

$$Z = \int \prod_{k<\Lambda} \frac{d\Phi^*(\vec{k})d\Phi(\vec{k})}{2\pi i} e^{-(S_{0<}+S_{0>}+S_{int})},\tag{3.12}$$

with

$$S_{0<} = \int_0^{\Lambda/b} \frac{d\vec{k}}{(2\pi)^d}(k^2 - \mu)|\Phi(\vec{k})|^2,\tag{3.13}$$

$$S_{0>} = \int_{\Lambda/b}^{\Lambda} \frac{d\vec{k}}{(2\pi)^d}(k^2 - \mu)|\Phi(\vec{k})|^2,\tag{3.14}$$

and

$$S_{int} = \lambda \int_0^{\Lambda} \frac{d\vec{k}_1 \dots d\vec{k}_4}{(2\pi)^{3d}} \delta(\vec{k}_1 + \vec{k}_2 - \vec{k}_3 - \vec{k}_4)\Phi^*(\vec{k}_4)\Phi^*(\vec{k}_3)\Phi(\vec{k}_2)\Phi(\vec{k}_1).\tag{3.15}$$

We want to integrate over the Fourier components with $\Lambda/b < k < \Lambda$ and try to manipulate the action for the remaining modes into the old form. The strategy will be perturbation theory in λ. To the zeroth order in λ we thus obtain $Z = Z_{0<}Z_{0>}$ since the partition function factorizes into a product of partition functions for slow and fast modes. Since in $Z_{0>}$ all the modes have large k, it contributes only to the regular, analytic part of the free energy. $Z_{0<}$, on the other hand, has the same form as the full non-interacting partition function, except that Λ is now replaced by Λ/b. To bring it into the exact old form to zeroth order in λ we first rescale the wavevectors as $b\vec{k} \to \vec{k}$, after which $S_{0<}$ becomes

$$S_{0<} = \frac{1}{b^d} \int_0^{\Lambda} \frac{d\vec{k}}{(2\pi)^d} \left(\frac{k^2}{b^2} - \mu \right) |\Phi(\vec{k})|^2. \tag{3.16}$$

Defining $\mu(b) = \mu b^2$, and rescaling $\Phi(\vec{k})/b^{(d+2)/2} \to \Phi(\vec{k})$, $S_{0<}$ in terms of the rescaled Fourier components now takes precisely the old form. Since the variables $\Phi(\vec{k})$ ultimately are to be integrated over, the critical behavior is unaffected by their rescaling: rescaling $\Phi(\vec{k})$ in the integration measure in Z only contributes an additive regular part to the free energy. So to the zeroth order in interaction we find

$$\mu(b) = \mu b^2 + O(\lambda), \tag{3.17}$$

and consequently by Eq. (3.9), $\nu = 1/2$. Since the susceptibility to zeroth order in λ in the normal phase is then

$$\chi(k) = \frac{b^2}{(bk)^2 - \mu(b)}, \tag{3.18}$$

it follows from Eq. (3.10) that $\gamma = 1$. The zeroth order calculation therefore simply recovers the familiar mean-field exponents.

Of course, the whole point of the exercise is to proceed beyond the mean-field approximation, which by avoiding the slow modes can now be done without encountering the infrared singularity. To first order in λ the partition function can be written as

$$Z = Z_{0>} \int \prod_{k<\Lambda/b} \frac{d\Phi^*(\vec{k})d\Phi(\vec{k})}{2\pi i} e^{-S_{0<}} \left[1 - \lambda \int_0^{\Lambda} \frac{d\vec{k}_1 \dots d\vec{k}_4}{(2\pi)^{3d}} \right.$$

$$\left. \times \delta(\vec{k}_1 + \vec{k}_2 - \vec{k}_3 - \vec{k}_4)\langle\Phi^*(\vec{k}_4)\Phi^*(\vec{k}_3)\Phi(\vec{k}_2)\Phi(\vec{k}_1)\rangle_{0>} + O(\lambda^2) \right]. \tag{3.19}$$

Figure 3.1 Diagrammatic representation of the first-order contribution to the renormalized chemical potential $\mu(b)$.

The average over the fast modes is defined as

$$\langle A \rangle_{0>} = \frac{1}{Z_{0>}} \int \prod_{\Lambda/b < k < \Lambda} \frac{d\Phi^*(\vec{k}) d\Phi(\vec{k})}{2\pi i} e^{-S_{0>}} A. \qquad (3.20)$$

Since the action S_0 is quadratic in Φ and diagonal in $\vec{k}$ we have that

$$\langle \Phi^*(\vec{k}_1) \Phi(\vec{k}_2) \rangle_0 = \frac{(2\pi)^d}{k_1^2 - \mu} \delta(\vec{k}_1 - \vec{k}_2), \qquad (3.21)$$

$$\langle \Phi^*(\vec{k}) \rangle_0 = \langle \Phi^*(\vec{k}_1) \Phi^*(\vec{k}_2) \rangle_0 = \langle \Phi^*(\vec{k}_1) \Phi(\vec{k}_2) \Phi(\vec{k}_3) \rangle_0 = 0, \qquad (3.22)$$

and

$$\langle \Phi^*(\vec{k}_4) \Phi^*(\vec{k}_3) \Phi(\vec{k}_2) \Phi(\vec{k}_1) \rangle_0 = \langle \Phi^*(\vec{k}_4) \Phi(\vec{k}_2) \rangle_0 \langle \Phi^*(\vec{k}_3) \Phi(\vec{k}_1) \rangle_0$$
$$+ (\vec{k}_1 \longleftrightarrow \vec{k}_2). \qquad (3.23)$$

The last three equations are simple properties of Gaussian integrals, such as $Z_{0>}$, as can be easily checked. Any combination containing a different number of Φs and Φ^*s averages to zero; otherwise the average factorizes into a product of all possible averages of pairs of Φ^* and Φ. This statement is also known as *Wick's theorem*.

The only combinations in Eq. (3.19) that yield a finite average therefore are when: (1) $\vec{k}_4 = \vec{k}_2$ and $\Lambda/b < k_2 < \Lambda$ and $\vec{k}_3 = \vec{k}_1$ and $k_1 < \Lambda/b$, or $\vec{k}_4 = \vec{k}_1$ and $\Lambda/b < k_1 < \Lambda$ and $\vec{k}_3 = \vec{k}_2$ and $k_2 < \Lambda/b$, (2) the same with $\vec{k}_4$ and $\vec{k}_3$ exchanged, and (3) when all $k < \Lambda/b$, (4) when all $\Lambda/b < k < \Lambda$. The last combination does not contribute to the action for the slow modes but only to the prefactor in Eq. (3.19) and thus to the analytic part of the free energy. We can represent the combinations (1) and (2) by the "tadpole" diagram in Fig, 3.1. Think of a quartic vertex as in Fig. 2.4: in combinations (1) and (2) two of the legs, one with outgoing and the other with ingoing arrow, are the fast modes to be integrated over, while the remaining two are the slow modes. The legs joined together in Fig. 3.1 represent the average of a pair of fast modes, and those left free, the remaining slow modes. The diagram in Fig 3.1 can thus be formed in four different ways.

Omitting the factor from the fast modes that contributes only to the analytic part of the free energy, we can write $Z \propto Z_<$ with the action in $Z_<$

$$S_< = \int_0^{\Lambda/b} \frac{d\vec{k}}{(2\pi)^d} \left(k^2 - \mu + 4\lambda \int_{\Lambda/b}^{\Lambda} \frac{d\vec{q}}{(2\pi)^d} \frac{1}{q^2 - \mu} \right) |\Phi(\vec{k})|^2$$

$$+ \lambda \int_0^{\Lambda/b} \frac{d\vec{k}_1 ... d\vec{k}_4}{(2\pi)^{3d}} \delta(\vec{k}_1 + \vec{k}_2 - \vec{k}_3 - \vec{k}_4) \Phi^*(\vec{k}_4) \Phi^*(\vec{k}_3) \Phi(\vec{k}_2) \Phi(\vec{k}_1)$$

$$+ O(\lambda^2). \tag{3.24}$$

Again we rescale the momenta to bring the cutoff back to Λ, and then rescale the slow Fourier components as described just below Eq. (3.16). After this we find

$$\mu(b) = b^2 \left(\mu - 4\lambda \int_{\Lambda/b}^{\Lambda} \frac{d\vec{q}}{(2\pi)^d} \frac{1}{q^2 - \mu} + O(\lambda^2) \right), \tag{3.25}$$

and

$$\lambda(b) = b^{4-d}\lambda + O(\lambda^2). \tag{3.26}$$

Let us analyze the result in Eq. (3.25) first. Due to the interaction the critical value of μ has been shifted to

$$\mu_c(b) = 4\lambda \int_{\Lambda/b}^{\Lambda} \frac{d\vec{q}}{(2\pi)^d} \frac{1}{q^2} + O(\lambda^2), \tag{3.27}$$

and become a function of b. This only means that, once all modes have been integrated out, the location of the critical point will depend on the cutoff Λ, and is therefore a non-universal quantity. This is expected: multiplied by the Boltzmann constant the transition temperature has the units of interaction and in principle depends on the microscopic details in the system. Let us then measure the chemical potential from its critical value at the current value of b: $\tilde{\mu} = \mu - \mu_c(b)$. Then

$$\tilde{\mu}(b) = b^2 \left(\tilde{\mu} - 4\lambda \int_{\Lambda/b}^{\Lambda} \frac{d\vec{q}}{(2\pi)^d} \left(\frac{1}{q^2 - \tilde{\mu}} - \frac{1}{q^2} \right) + O(\lambda^2) \right). \tag{3.28}$$

In the critical region when $|\tilde{\mu}| \ll \Lambda^2$ this simplifies into

$$\tilde{\mu}(b) = b^2 \tilde{\mu} \left(1 - 4\lambda \frac{\Lambda^{d-4} S_d}{(2\pi)^d} \int_{1/b}^1 \frac{x^{d-1} dx}{x^4} + O(\lambda^2) \right), \tag{3.29}$$

where $S_d = 2\pi^{d/2}/\Gamma(d/2)$ is the area of the unit sphere in d dimensions, and $\Gamma(z)$ is the factorial function. The integral is

$$\int_{1/b}^{1} \frac{x^{d-1}dx}{x^4} = \frac{1}{d-4}\left(1 - \frac{1}{b^{d-4}}\right) = \ln b + O(d-4), \qquad (3.30)$$

assuming that $|d - 4| \ll 1$. Defining the dimensionless interaction as $\hat{\lambda} = \lambda \Lambda^{d-4} S_d/(2\pi)^d$, we finally write

$$\tilde{\mu}(b) = \tilde{\mu}b^2(1 - 4\hat{\lambda}\ln(b) + O(\hat{\lambda}(d-4), \hat{\lambda}^2, \hat{\lambda}\tilde{\mu})) \approx \tilde{\mu}b^{2-4\hat{\lambda}+O(\hat{\lambda}^2)}. \quad (3.31)$$

Assuming $\hat{\lambda}$ to be small, there are therefore three possibilities for the chemical potential as $b \to \infty$.

(1) For $\tilde{\mu} > 0$, $\tilde{\mu}(b) \to \infty$. This limit corresponds to the superfluid (broken symmetry) phase.
(2) When $\tilde{\mu} < 0$, $\tilde{\mu}(b) \to -\infty$. This corresponds to the normal (symmetric) phase.
(3) When $\tilde{\mu} = 0$, that is when $\mu = \mu_c(b)$, $\tilde{\mu}(b) \equiv 0$. This is the *fixed point* of the transformation. From Eq. (3.31) it follows that this is an *unstable* fixed point, and that $\tilde{\mu}$ grows when perturbed from its fixed point value. This is correct for both zero and finite $\hat{\lambda}$, as long as it is not too large. The variable $\tilde{\mu}$, which is the tuning parameter for the transition, is thus a relevant coupling.

From the definition of y above Eq. (3.8) and Eq. (3.31) we see that $y = 2(1 - 2\hat{\lambda}^* + O((\hat{\lambda}^*)^2))$. By Eq. (3.9) then,

$$\nu = \frac{1}{2} + \hat{\lambda}^* + O((\hat{\lambda}^*)^2), \qquad (3.32)$$

where $\hat{\lambda}^* = \lim_{b \to \infty} \hat{\lambda}$. On the other hand, Eq. (3.26) implies that for $d > 4$ and $\hat{\lambda} \ll 1$, $\hat{\lambda}^* = 0$. Above four dimensions weak interaction scales to zero under the momentum-shell transformation, and it is irrelevant. Once again we recover the mean field exponents as exact. Note that $\hat{\lambda}^* = 0$ represents a stable fixed point above four dimensions. When $d < 4$, on the other hand, a weak interaction grows with b and the next-order term must be included. For $d < 4$ then both the chemical potential and the interaction are relevant couplings at the non-interacting fixed point $\hat{\lambda}^* = 0$.

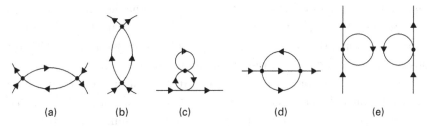

Figure 3.2 Second-order diagrams. Diagrams (a) and (b) contribute to the renormalized interaction, and (c) and (d) to the quadratic part of $S_<$. Diagram (e) is disconnected and therefore non-contributing.

Figure 3.3 Two more non-contributing second-order diagrams.

To find $\hat{\lambda}^*$ below four dimensions we need to compute the $O(\lambda^2)$ term in Eq. (3.26). We start by explicitly writing the $O(\lambda^2)$ term in Eq. (3.19):

$$\frac{\lambda^2}{2} \int_0^\Lambda \frac{d\vec{p}_1...d\vec{p}_4 d\vec{q}_1...d\vec{q}_4}{(2\pi)^{6d}} \delta(\vec{p}_1 + \vec{p}_2 - \vec{p}_3 - \vec{p}_4)\delta(\vec{q}_1 + \vec{q}_2 - \vec{q}_3 - \vec{q}_4)$$
$$\times \langle \Phi^*(\vec{p}_4)\Phi^*(\vec{p}_3)\Phi(\vec{p}_2)\Phi(\vec{p}_1)\Phi^*(\vec{q}_4)\Phi^*(\vec{q}_3)\Phi(\vec{q}_2)\Phi(\vec{q}_1)\rangle_{0>}. \qquad (3.33)$$

In computing the above average we need to pair up Φ^*s with Φs with equal wavevectors $\Lambda/b < k < \Lambda$ in all possible ways. In the language of diagrams, we need to join the legs of two vertices like in Fig. 2.4, keeping in mind that a leg with an arrow going out may be joined only to a leg with an arrow going in. This produces the diagrams in Fig. 3.2. Diagrams where the number of unpaired legs is fewer than two or more than four are omitted, since these do not contribute to either $\mu(b)$ or $\lambda(b)$. Diagrams that have been depicted separately in Fig. 3.3 do not contribute either, because conservation of momentum implies that when all the external legs carry a low momentum the horizontal internal line cannot carry a large momentum. The diagrams in Fig. 3.3 are also called *one-particle reducible* in the literature on field theory.

The diagrams in Fig. 3.2 (c) and (d) have only two external legs and thus contribute to the quadratic part of $S_<$. Since the lowest-order contribution to the quadratic part was already finite, to the first order in $\hat{\lambda}^*$ they may be neglected. The diagram in Fig. 3.2 (e) has four external legs and looks like it could potentially contribute to $\lambda(b)$. It is, however, in contrast to the diagrams in Fig. 3.2 (a) and (b), *disconnected*. The reader is invited to check that this causes it to cancel out upon re-exponentiation into $S_<$. One can show that this rule holds to all orders in λ, and in our calculation we therefore need to consider only the *connected* diagrams. This is an example of the *linked-cluster theorem*, an economical proof of which is presented in Appendix B. To quadratic order in λ, we are left then only with the diagrams in Fig. 3.2 (a) and (b). The diagram in Fig. 3.2 (a) can be constructed in sixteen different ways, and the one in Fig. 3.2 (b) in four. Upon re-exponentiation the $O(\lambda^2)$ term in $S_<$ becomes

$$
-\frac{\lambda^2}{2} \int_0^{\Lambda/b} \frac{d\vec{k}_1...d\vec{k}_4}{(2\pi)^{3d}} \delta(\vec{k}_1 + \vec{k}_2 - \vec{k}_3 - \vec{k}_4)\Phi^*(\vec{k}_4)\Phi^*(\vec{k}_3)\Phi(\vec{k}_2)\Phi(\vec{k}_1)
$$
$$
\times \left[16 \int_{\Lambda/b}^{\Lambda} \frac{d\vec{q}}{(2\pi)^d} \frac{1}{(q^2 - \tilde{\mu})((\vec{q} + \vec{k}_2 - \vec{k}_4)^2 - \tilde{\mu})} \right.
$$
$$
\left. + 4 \int_{\Lambda/b}^{\Lambda} \frac{d\vec{q}}{(2\pi)^d} \frac{1}{(q^2 - \tilde{\mu})((\vec{k}_1 + \vec{k}_2 - \vec{q})^2 - \tilde{\mu})} \right], \tag{3.34}
$$

where we have also replaced μ with $\tilde{\mu}$. Since the difference between the two is $O(\lambda)$ this only affects the next-order term.

Although the original interaction in Eq. (3.15) was the delta-function in real space and thus independent of the fields' wavevectors, the momentum-shell transformation generated some wavevector dependence in the interaction for the slow modes, as evident from the last equation. Although not obvious at this point, it will prove possible to neglect this dependence and still define a single interaction coupling constant $\lambda(b)$ as the value of the new interaction at a particular value of the momentum. For convenience we will choose that value to be at $\vec{k}_i = 0$. After the standard rescaling of the wavevectors and the fields, we then find that

$$
\lambda(b) = b^{4-d}\lambda \left(1 - 10\lambda \int_{\Lambda/b}^{\Lambda} \frac{d\vec{q}}{(2\pi)^d} \frac{1}{(q^2 - \tilde{\mu})^2} + O(\lambda^2) \right). \tag{3.35}
$$

At the critical point $\tilde{\mu} = 0$, after performing the integral and introducing $\hat{\lambda}$, we finally find

$$
\hat{\lambda}(b) = b^{4-d}\hat{\lambda}(1 - 10\hat{\lambda} \ln b + O(\hat{\lambda}^2)) = b^{4-d-10\hat{\lambda}+O(\hat{\lambda}^2)}\hat{\lambda}. \tag{3.36}
$$

Together with the result in Eq. (3.31) this completes the calculation to the lowest non-trivial order. In the next section we will discuss how the

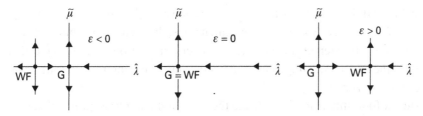

Figure 3.4 The flow of the renormalized interaction and the chemical potential above, at, and below four dimensions. The Wilson–Fisher fixed point moves from the unphysical region for $\epsilon < 0$ to the physical region for $\epsilon > 0$, and exchanges stability along the λ-direction with the Gaussian fixed point.

functions $\tilde{\mu}(b)$ and $\hat{\lambda}(b)$ may be used to systematically compute the critical exponents.

3.3 ε-expansion

It is convenient to cast first the derived momentum-shell transformations $\mu(b)$ and $\lambda(b)$ into differential form. From Eqs. (3.31) and (3.36), for $|4 - d| \ll 1$ we can write

$$\beta_\mu \equiv \frac{d\tilde{\mu}(b)}{d \ln b} = \tilde{\mu}(b)(2 - 4\hat{\lambda}(b) + O(\epsilon\hat{\lambda}, \hat{\lambda}^2, \hat{\lambda}\tilde{\mu})), \qquad (3.37)$$

$$\beta_\lambda \equiv \frac{d\hat{\lambda}(b)}{d \ln b} = \epsilon\hat{\lambda}(b) - 10\hat{\lambda}^2(b) + O(\epsilon\hat{\lambda}^2, \hat{\lambda}^3, \tilde{\mu}\hat{\lambda}^2), \qquad (3.38)$$

where $\epsilon = 4 - d$. The momentum-shell transformation this way may be understood as a flow in the space of coupling constants generated by the above differential equations. The derivatives of the couplings with respect to parameter b are called *β-functions*. Three different scenarios are depicted in Fig. 3.4.

The easiest way to understand qualitatively the flow in the $\hat{\lambda} - \tilde{\mu}$ plane is to look for the fixed points at which $\beta_\mu = \beta_\lambda = 0$ and analyze their stability. The *differential recursion relations* in Eqs. (3.37) and (3.38) have two such fixed points: the Gaussian at $\hat{\lambda} = 0$, and $\tilde{\mu} = 0$, and the Wilson–Fisher fixed point at $\hat{\lambda} = (\epsilon/10) + O(\epsilon^2)$ and $\tilde{\mu} = 0$. Above four dimensions $\epsilon < 0$ and the Wilson–Fisher fixed point is at a negative interaction and therefore unphysical. It is also unstable in both λ- and μ-directions, i.e. it has two relevant couplings. The Gaussian fixed point, on the other hand, is stable in the λ-direction and unstable only in the μ-direction. Such a fixed point with precisely one unstable direction will be called a *critical point*. The unstable line $\tilde{\mu} = 0$ is an example of a *critical surface*. So, above four dimensions, the mean-field approximation which neglects interactions is exact in the critical region because the critical point is the non-interacting Gaussian fixed point: small interaction is irrelevant and flows to zero.

Below four dimensions $\epsilon > 0$ and the Wilson–Fisher fixed point moves into the physical region and exchanges stability with the Gaussian fixed point: the Wilson–Fisher fixed point becomes stable in the λ direction while the Gaussian one becomes completely unstable. Critical behavior for $\epsilon > 0$ is governed by the non-trivial Wilson–Fisher fixed point, the location of which is determined by ϵ. The parameter ϵ controls the size of corrections to the mean-field critical behavior. It is therefore useful to consider ϵ as a continuous parameter, and in particular to assume that $\epsilon \ll 1$ so as to exert some control over the magnitude of the Wilson–Fisher fixed point.

In $d = 4$ the Wilson–Fisher and the Gaussian fixed points happen to coincide. A weak positive interaction flows to zero, but since the linear term in β_λ vanishes the flow is slow. This leads to the so-called logarithmic corrections to scaling right at the upper critical dimension (Problem 3.4).

Flow near the fixed points follows from the *stability matrix M*:

$$M_{ij} \equiv \frac{\partial \beta_{x_i}}{\partial x_j}\Big|_{x_i = x_i^*},\qquad (3.39)$$

where $x_i = \tilde{\mu}, \hat{\lambda}$ in the present case. The eigenvectors of M with positive eigenvalues determine the relevant (unstable) directions at the given fixed point, and those with negative eigenvalues the irrelevant (stable) directions. In the case of a critical point there is only one such relevant direction, with the eigenvalue y. The correlation length exponent is then given by $\nu = 1/y$. In the present case M is already diagonal at the Wilson–Fisher fixed point with $y = 2 - 2\epsilon/5$. The correlation length exponent is therefore

$$\nu = \frac{1}{2} + \frac{\epsilon}{10} + O(\epsilon^2).\qquad (3.40)$$

For $\epsilon = 1$ this leads to $\nu \approx 0.6$ in three dimensions, encouragingly closer to the experimental value $\nu = 0.670$.

The second eigenvalue of M at the Wilson–Fisher fixed point is $-\epsilon$ and determines how rapidly the interaction approaches the fixed point value. It determines the *corrections to scaling*, which will be discussed in Section 3.5.

Since a single eigenvalue y determines the flow for both a slightly positive and a slightly negative chemical potential near the critical point, the correlation length exponent ν is equal on both sides of the transition. What are the values of the other exponents? Let us find the value of the anomalous dimension η first. We can express the result of the momentum-shell transformation, before the rescaling of the momenta and of the fields, by the change in the action for slow modes, as

$$k^2 \rightarrow Z_k k^2, \tag{3.41}$$

$$\mu \rightarrow Z_\mu \mu, \tag{3.42}$$

and

$$\lambda \rightarrow Z_\lambda \lambda. \tag{3.43}$$

To the lowest order and near four dimensions the renormalization factors are found to be $Z_k = 1 + O(\hat{\lambda}^2)$, $Z_\mu = 1 - 4\hat{\lambda} \ln b + O(\hat{\lambda}^2)$, and $Z_\lambda = 1 - 10\hat{\lambda} \ln b + O(\hat{\lambda}^2)$. The anomalous dimension is defined as

$$\eta = \frac{dZ_k}{d \ln b}\big|_{\hat{\lambda}=\hat{\lambda}^*}. \tag{3.44}$$

From the lowest-order result for Z_k it then follows that $\eta = O(\epsilon^2)$. Let us nevertheless repeat our rescaling procedure assuming a finite η. To bring the cutoff back to Λ, momenta are still rescaled as $bk \rightarrow k$ as before, but rescaling of the fields needs to be modified into $\Phi/b^{(2+d-\eta)/2} \rightarrow \Phi$. When $\eta \neq 0$, then

$$\tilde{\mu}(b) = b^{2-\eta} \tilde{\mu} Z_\mu \tag{3.45}$$

and

$$\hat{\lambda}(b) = b^{\epsilon - 2\eta} \hat{\lambda} Z_\lambda. \tag{3.46}$$

The "sunrise" diagram in Fig. 3.2 (d) leads to the contribution $O(\epsilon^2)$ in η. The calculation of this diagram is quite involved due to the fact that there are three internal lines, and we relegate it to Section 3.7 where a more powerful field-theoretic formulation of renormalization group will be used. Let us here assume only that for $|\epsilon| \ll 1$ it yields

$$Z_k = 1 + c\hat{\lambda}^2 \ln b + O(\hat{\lambda}^3) \approx b^{c\hat{\lambda}^2}, \tag{3.47}$$

where c is a number that depends on the number of the field's components. The anomalous dimension by Eq. (3.44) is then $\eta = c(\hat{\lambda}^*)^2$.

When $\eta \neq 0$, the superfluid susceptibility for the slow modes with $k < \Lambda/b$ in the normal phase is then

$$\chi(k) = \frac{b^{2-\eta}}{(bk)^2 - \tilde{\mu}(b)}, \tag{3.48}$$

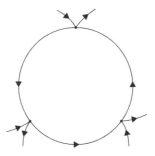

Figure 3.5 The diagram generating the $|\Phi_<|^6$ term.

which has the form as in Eq. (3.4), with $x = 2 - \eta$. Since $\nu = 1/y$ we see that
Eq. (3.10) is nothing but Fisher's scaling law. Computing η from Eq. (3.44)
is therefore in accord with its definition in Eq. (1.26).

Let us finally discuss the fate of the new coupling constants in $S_<$ generated
by the momentum-shell transformation. A good example is provided by the
sixth-order term $g|\Phi|^6$, which, although absent initially, becomes generated
by the diagram in Fig. 3.5. The standard rescaling of the fields implies that

$$\beta_g = \frac{dg}{d\ln b} = 2(3 - d)g + O(g\hat{\lambda}), \tag{3.49}$$

and a weak coupling g is irrelevant at the Wilson–Fisher fixed point near
$d = 4$. Right at $d = 3$, however, the linear term vanishes, and g becomes a
marginal coupling at the Gaussian fixed point. It is believed, however, that the
higher order terms in β_g still make g *irrelevant* at the Wilson–Fisher fixed point
in $d = 3$ (see Problem 3.3). Simple dimensional analysis shows that all other
couplings, like those in Eq. (3.7) for example, are also irrelevant near $d = 4$. In
particular, this is what allows one to neglect the generated wavevector dependence of the interaction in Eq. (3.34), and justifies the approximation of the
physical interaction between helium atoms with the delta-function repulsion
accompanied by the ultraviolet cutoff.

Problem 3.1 Find the correlation length exponent ν to the lowest order in
ϵ when Φ has N real components, and use the result to estimate it in two
and three dimensions for the phase transitions in the Ising universality class
($N = 1$).

Solution For M complex components we first generalize the interaction term
in the action into $(\sum_{i=1}^{M} |\Phi_i|^2)^2$. The diagrams in Figs. 3.1, and 3.2 (a) and

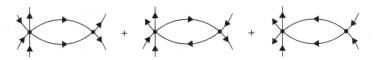

Figure 3.6 Diagrams contributing to the $O(g\hat{\lambda})$ term in Eq. (3.49).

(b) then lead to

$$\beta_\mu = \mu(2 - 2(M + 1)\hat{\lambda})$$

and

$$\beta_\lambda = \epsilon\hat{\lambda} - 2(M + 4)\hat{\lambda}^2.$$

Taking $M = N/2$, this implies

$$\nu = \frac{1}{2} + \frac{N + 2}{4(N + 8)}\epsilon + O(\epsilon^2).$$

For $N = 1$ the lowest order estimate is therefore $\nu \approx 0.58$ in $d = 3$, and $\nu \approx 0.66$ in $d = 2$. This should be compared with the almost exact value $\nu = 0.63$ from high-temperature expansion in three dimensions, and with $\nu = 1$ from Onsager's exact solution in two dimensions.

Problem 3.2 Find the value of the anomalous dimension at the superfluid transition in ^{4}He if $\eta = \epsilon^2/54$ at the Ising ferromagnetic transition.

Solution The sunrise diagram in Fig. 3.2 (d) can be formed in $8(M + 1)$ ways when the field has M complex components. We can thus write $\eta = c8(M + 1)(\lambda^*)^2$, with $\lambda^* = \epsilon/(2(M + 4))$, and c as a constant. Taking $M = N/2$, the given value at $N = 1$ then fixes $c = 1/8$, and

$$\eta = \frac{N + 2}{2(N + 8)^2}\epsilon^2 + O(\epsilon^3).$$

For $N = 2$ then $\eta = \epsilon^2/50$.

Problem 3.3 Determine the relevance of the sixth-order term's coupling g at the Wilson–Fisher critical point to the lowest order in ϵ.

Solution The diagrams in Fig. 3.6 can be formed in $6(M + 6)$ ways when the field has M complex components. Therefore

$$\beta_g = 2[(3 - d) - 3(M + 6)\lambda + O(\lambda^2)]g + O(g^2),$$

so precisely in $d = 3$, $\beta_g \approx (-3\epsilon(M+6)/(M+4) + O(\epsilon^2))g$. This suggests that g is irrelevant at the Wilson–Fisher fixed point in $d = 3$. The validity of this conclusion, however, obviously depends on the higher-order terms in the expansion.

3.4 Dangerously irrelevant coupling

To find the remaining exponents α, β, and δ, and thus check the validity of the scaling laws, we first couple a weak external source j to the order parameter, like in Eq. (2.42), assuming $j(\vec{r}) = j$ constant and real. Similarly to the susceptibility in Eq. (3.6), the free energy per unit volume at the cutoff Λ/b in the superfluid phase may then be written as

$$F = b^{-d} F_-(\mu(b), \lambda(b), j(b)). \qquad (3.50)$$

Since the term $\sim j(\Phi + \Phi^*)$ breaks the $U(1)$ symmetry of the action, it cannot become generated by the integration over the fast modes. This implies that the transformation $j(b)$ at small j is determined exactly by the rescaling of the field as given immediately before Eq. (3.45): $j(b) = jb^{(2+d-\eta)/2}$. The source, or the magnetic field in the case of magnetic transitions, is therefore another relevant coupling, both at the Gaussian and the Wilson–Fisher fixed points. This was to be expected, since, just like the chemical potential, the source also needs to be finely tuned for the critical point to be reached.

Near the critical point, $\mu(b) = b^{1/\nu}\mu$, and taking again $\mu(b) = \mu_0$ at $b \gg 1$, we obtain

$$F = \left(\frac{\mu}{\mu_0}\right)^{\nu d} F_-\left(\mu_0, \lambda^*, j\left(\frac{\mu}{\mu_0}\right)^{-\nu(2+d-\eta)/2}\right), \qquad (3.51)$$

which indeed has the scaling form in Eq. (1.9). Differentiating with respect to j we find

$$\langle \Phi \rangle = -\left(\frac{\mu}{\mu_0}\right)^\beta \frac{\partial F_-(\mu_0, \lambda^*, j)}{\partial j}\Big|_{j=0}, \qquad (3.52)$$

and after recalling that $\mu \propto (T_c - T)$,

$$\beta = \frac{\nu(d + \eta - 2)}{2}. \qquad (3.53)$$

Setting $j = 0$ and differentiating the free energy twice with respect to μ one similarly finds Josephson's scaling law, and the specific heat exponent

$\alpha = 2 - \nu d$. Eliminating d in Eq. (3.53) and using Fisher's law implies then Rushbrooke's scaling law.

At the transition point $\mu = 0$ one can alternatively choose the factor b so that $jb^{(2+d-\eta)/2} = j_0$, with j_0 constant. Then

$$F = \left(\frac{j}{j_0}\right)^{\frac{2d}{2+d-\eta}} F_-(0, \lambda^*, j_0). \tag{3.54}$$

Differentiating with respect to j, it follows that

$$\delta = \frac{d + 2 - \eta}{d - 2 + \eta}. \tag{3.55}$$

Combining with Eq. (3.53) and Josephson's law, this leads to Griffiths' scaling law in Eq. (1.23).

It therefore seems that all the scaling laws are satisfied. However, above $d = 4$ where $\nu = 1/2$ and $\eta = 0$, our result is $\beta = (d - 2)/4 \neq 1/2$ and $\delta = (d + 2)/(d - 2) \neq 3$, and the obtained values do not agree with the mean-field theory. What went wrong? The problem lies in the value of λ^*: for $d < 4$, $\lambda^* > 0$ and it is legitimate to replace $\lambda(b)$ with λ^* in the free energy, since the free energy is an analytic function of λ at a finite $\lambda = \lambda^*$. For $d > 4$, on the other hand, $\lambda^* = 0$, and the free energy is non-analytic at $\lambda = \lambda^*$. Recall the result of Landau's mean-field theory: in the ordered phase $F \sim 1/\lambda$, as in Eq. (2.39), for example. More generally, one can see that the free energy is non-analytic at $\lambda = 0$ by noticing that for an infinitesimal $\lambda < 0$ the action would become unbounded from below, and the free energy infinite. The radius of convergence of the perturbation series for free energy in λ therefore must be zero, and one cannot simply replace $\lambda(b)$ with λ^* when $\lambda^* = 0$. The interaction is an example of a *dangerously irrelevant coupling* in dimensions $d > 4$. Consequently, the hyperscaling assumption is violated above the upper critical dimension.

3.5 Corrections to scaling

Let us assume that $d < 4$ so that $\lambda^* > 0$, and reconsider the scaling form for the normal phase susceptibility in Eq. (3.4). For a large b we can expand the scaling function near the fixed point value and write

$$\chi = b^x \left[F(bk, |\mu(b)|, \lambda^*) + (\lambda(b) - \lambda^*) \frac{\partial F(bk, |\mu(b)|, \lambda)}{\partial \lambda} \Big|_{\lambda = \lambda^*} \right.$$
$$\left. + O((\lambda(b) - \lambda^*)^2) \right]. \tag{3.56}$$

Choosing again $\mu(b) = \mu_0$, and using the fact that $\lambda(b) - \lambda^* = (\lambda - \lambda^*)b^{-\epsilon}$ to the lowest order in ϵ,

$$
\chi = \left|\frac{\mu_0}{\mu}\right|^{\gamma} \left[F\left(k\left|\frac{\mu_0}{\mu}\right|^{\nu}, |\mu_0|, \lambda^*\right) \right.
$$

$$
\left. + (\lambda - \lambda^*)\frac{\partial F(k|\frac{\mu_0}{\mu}|^{\nu}, |\mu_0|, \lambda^*)}{\partial \lambda}\left|\frac{\mu}{\mu_0}\right|^{\epsilon/2 + O(\epsilon^2)} + O\left((\lambda(b) - \lambda^*)^2\right) \right].
$$

$$(3.57)$$

The first term is the familiar scaling form of the susceptibility. The second term provides the *correction to scaling*. For $\epsilon \ll 1$ the correction coming from the interaction is the *leading correction*, since the power $\epsilon/2 + O(\epsilon^2)$ is arbitrarily small. The other *subleading* corrections come from the more irrelevant couplings which have already been set to zero in χ, like the sixth-order term coupling g, for example. The leading correction to scaling is therefore determined by the least irrelevant coupling at the critical point, which in our case and near four dimensions is just the interaction measured from its fixed point value, $\lambda - \lambda^*$. The corrections to scaling need to be taken into account when extracting values of the critical exponents from experiment.[1] This also shows why only close to the critical point can the simple scaling be observed: not before all irrelevant couplings settle to their fixed-point values will the simple scaling emerge.

Problem 3.4 Find the logarithmic correction to the scaling of the correlation length in $d = 4$.

Solution Precisely in $d = 4$, $d\ln|\mu|/d\ln b = 2(1 - (M + 1)\lambda)$ and $d\lambda/d\,lnb = -2(M + 4)\lambda^2$, where M is the number of complex components (Problem 3.1). The second equation gives

$$
\lambda(b) = \frac{\lambda}{1 + 2\lambda(M + 4)\ln b} \approx \frac{1}{2(M + 4)\ln b},
$$

when $b \gg 1$. When inserted into the first equation, for $b \gg 1$ one finds

$$
\frac{\mu(b)}{\mu} \approx \frac{b^2}{(\ln b)^{\frac{M+1}{M+4}}}.
$$

[1] For example, the quoted value of α at the superfluid transition in ^{4}He was obtained by taking $\epsilon = 1$ in the lowest-order result for the leading correction to scaling of the specific heat.

Since the correlation length $\xi \sim b$, taking $\mu(b) = \mu_0$ yields

$$\xi \propto \frac{(\ln |\frac{\mu_0}{\mu}|)^{\frac{M+1}{2(M+4)}}}{|\frac{\mu}{\mu_0}|^{\frac{1}{2}}}.$$

3.6 Field-theoretic perspective*

The renormalization group may also be understood as a way of summing certain classes of terms in the perturbation series to infinite order. This is often called the field-theoretic approach, and is of course ultimately completely equivalent to the momentum-shell procedure we have just described. It is, however, more computationally powerful if one wants to proceed beyond the first-order calculation, and also provides a conceptually different perspective on renormalization that is in some ways simpler. In this and the following section we discuss briefly only the gist of the method, sacrificing some rigor to avoid becoming overly formal.

Let us consider again the superfluid susceptibility $\chi(k)$ in the normal phase, and try to compute the exponent γ, for example, by using the direct perturbation theory in the interaction λ. The first-order contribution is then given by the same diagram in Fig. 3.1, except that the integration is performed over all wavevectors, not only over the large ones as done previously. After a calculation completely analogous to the one that led to Eq. (3.25) we find

$$\chi^{-1}(k) = k^2 + |\tilde{\mu}| \left(1 - 2(M+1)\lambda \int_0^\Lambda \frac{d\vec{q}}{(2\pi)^d} \frac{1}{q^2(q^2 + |\tilde{\mu}|)}\right) + O(\lambda^2),$$
(3.58)

with M as the number of complex components of the field. The exponent γ is then obtained as

$$\gamma = \lim_{|\tilde{\mu}| \to 0} \frac{d \ln \chi^{-1}(k = 0)}{d \ln |\tilde{\mu}|}.$$
(3.59)

Expanding $\ln \chi^{-1}$ in powers of interaction and then performing the differentiation gives

$$\gamma = 1 + 2(M+1)\frac{\lambda S_d}{(2\pi)^d} \lim_{|\tilde{\mu}| \to 0} |\tilde{\mu}|^{-\epsilon/2} \int_0^{\Lambda/|\tilde{\mu}|^{1/2}} \frac{x^{d-1}dx}{x^2(x^2 + 1)^2} + O(\lambda^2).$$
(3.60)

Since the integral appearing is finite in the limit $|\tilde{\mu}| \to 0$, we can write

$$\gamma = 1 + 2(M+1) \int_0^\infty \frac{x^{d-3}dx}{(x^2 + 1)^2} \lim_{|\tilde{\mu}| \to 0} \hat{\lambda} + O(\hat{\lambda}^2),$$
(3.61)

with

$$\hat{\lambda} = \frac{\lambda S_d}{(2\pi)^d} |\tilde{\mu}|^{-\epsilon/2} \tag{3.62}$$

again as the dimensionless interaction. For $\epsilon > 0$ we thus find

$$\gamma = \infty. \tag{3.63}$$

This is, of course, the expected result, and the infinity reflects the breakdown of the direct perturbation theory in the critical region. For $\epsilon < 0$, on the other hand, $\gamma = 1$, and the perturbation theory correctly gives the mean-field behavior.

The result may be rendered finite, however, by introducing the renormalized interaction λ_r as the value of the full interaction vertex at the vanishing external momenta. The lowest-order contribution to λ_r is again given by the diagrams in Fig. 3.2 (a) and (b):

$$\lambda_r = \lambda - \frac{\lambda^2}{2} 4(M+4) \int_0^\Lambda \frac{d\vec{q}}{(2\pi)^d} \frac{1}{(q^2 + |\tilde{\mu}|)^2} + O(\lambda^3). \tag{3.64}$$

Multiplying both sides by the factor $S_d |\tilde{\mu}|^{-\epsilon/2}/(2\pi)^d$, this implies

$$\hat{\lambda}_r = \hat{\lambda} - 2(M+4)\hat{\lambda}^2 \int_0^{\Lambda/|\tilde{\mu}|^{1/2}} \frac{x^{d-1} dx}{(x^2+1)^2} + O(\hat{\lambda}^3). \tag{3.65}$$

Since to the lowest order $\hat{\lambda}_r = \hat{\lambda}$, the exponent γ in Eq. (3.61) can also be written as

$$\gamma = 1 + 2(M+1) \int_0^\infty \frac{x^{d-3} dx}{(x^2+1)^2} \lim_{|\tilde{\mu}| \to 0} \hat{\lambda}_r + O(\hat{\lambda}_r^2), \tag{3.66}$$

to the lowest order in the *renormalized* interaction $\hat{\lambda}_r$.

We can now find the requisite limit of $\hat{\lambda}_r$ in the following way. First we evaluate the integral in Eq. (3.65) for $d < 4$:

$$\int_0^\infty \frac{x^{d-1} dx}{(x^2+1)^2} = \frac{1}{2} \Gamma\left(\frac{d}{2}\right) \Gamma\left(\frac{\epsilon}{2}\right) = \frac{1}{\epsilon} + O(1), \tag{3.67}$$

with the last equality holding near four dimensions. Then we use first two terms in the renormalized interaction to compute the β-function as

$$\beta(\hat{\lambda}_r) = \frac{d\hat{\lambda}_r}{d \ln(|\tilde{\mu}|^{-1/2})} = \epsilon\hat{\lambda} - 4(M+4)\epsilon\hat{\lambda}^2 \left(\frac{1}{\epsilon} + O(1)\right)$$

$$= \epsilon\hat{\lambda}_r - 2(M+4)\hat{\lambda}_r^2 + O\left(\epsilon\hat{\lambda}_r^2, \hat{\lambda}_r^3\right). \tag{3.68}$$

Figure 3.7 "Parquet" diagrams summed by the lowest-order β-function in four dimensions.

From the β-function we see that

$$\lim_{|\tilde{\mu}|\to 0} \hat{\lambda}_r = \frac{\epsilon}{2(M+4)} + O(\epsilon^2), \qquad (3.69)$$

and thus

$$\gamma = 1 + \frac{M+1}{2(M+4)}\epsilon + O(\epsilon^2), \qquad (3.70)$$

in agreement with the result of the momentum-shell renormalization group.

The reader may be excused if feeling somewhat bewildered by what just happened. After all, although it is legitimate to replace the series in $\hat{\lambda}$ for the one in $\hat{\lambda}_r$ before the limit is taken in Eq. (3.61), $\hat{\lambda}_r$ from Eq. (3.65) looks just as infinite in the critical region as the original interaction $\hat{\lambda}$. The trick is that we obtained the critical value of $\hat{\lambda}_r$ not directly from Eq. (3.65), but from the β-function in Eq. (3.68). The β-function contains more information than just the first two terms in the perturbation theory for $\hat{\lambda}_r$. Consider the solution of the differential equation in Eq. (3.68):

$$\hat{\lambda}_r = \frac{\hat{\lambda}_{r0}}{1 + \frac{2(M+4)\hat{\lambda}_{r0}}{\epsilon}\left(\left(\frac{|\tilde{\mu}_0|}{|\tilde{\mu}|}\right)^{\epsilon/2} - 1\right)}\left(\frac{|\tilde{\mu}_0|}{|\tilde{\mu}|}\right)^{\epsilon/2}, \qquad (3.71)$$

where we assumed $\hat{\lambda}_{r0} < \epsilon/(2(M+4))$. Expanding in powers of $\hat{\lambda}_{r0}$ to second order and identifying $\lambda S_d/(2\pi)^d|\tilde{\mu}_0|^{\epsilon/2} = \hat{\lambda}_{r0}$ recovers the original Eq. (3.64). The full solution in the last equation, however, contains terms to infinite order in λ, and this is why the critical value of $\hat{\lambda}_r$ turns out finite.

Precisely in $d = 4$ the solution of the β-function becomes

$$\hat{\lambda}_r = \frac{\hat{\lambda}_{r0}}{1 + (M+4)\hat{\lambda}_{r0}\ln\left(\frac{|\tilde{\mu}_0|}{|\tilde{\mu}|}\right)}. \qquad (3.72)$$

When expanded this gives a series in which the coefficient of $\hat{\lambda}_{r0}^{n+1}$ is $\sim$ $[\ln(\frac{|\tilde{\mu}_0|}{|\tilde{\mu}|})]^n$, and is thus called the "leading logarithm" approximation. In effect, the result is the sum of the infinite series of diagrams in Fig. 3.7, often called the "parquet" diagrams, since they may be though of as formed by assembling together the lower-order diagrams in the series. The $\hat{\lambda}_r^3$ term in the β-function in $d = 4$ then adds the first subleading logarithms $\sim [\ln(\frac{|\tilde{\mu}_0|}{|\tilde{\mu}|})]^{n-1}$ to the coefficients in the series in λ (see Problem 3.5). The β-function may therefore be viewed as effectively summing whole classes of diagrams to infinite order in the coupling constant, arranged by the degree of their divergence in the critical region.

Finally, we should mention that although the ϵ-expansion was used in evaluating the integrals in Eqs. (3.61) and (3.66), this was not necessary. Sacrificing some control over the procedure, one could evaluate such integrals directly in $d = 3$, for example, since they are perfectly finite. This leads to the so-called fixed-dimension renormalization group, the results of which are cited in the Section 3.8. It turns out to be probably the most powerful analytic procedure for computing the critical behavior.

Problem 3.5* Assume that the renormalized interaction in some model at its upper critical dimension is given by the series

$$\lambda_r = \lambda + a\lambda^2 \ln\left(\frac{\Lambda}{|\tilde{\mu}|^{1/2}}\right) + \lambda^3\left[b\ln^2\left(\frac{\Lambda}{|\tilde{\mu}|^{1/2}}\right) + c\ln\left(\frac{\Lambda}{|\tilde{\mu}|^{1/2}}\right)\right] + O(\lambda^4),$$

where a, b, c are some finite coefficients, Λ the ultraviolet cutoff, and $\tilde{\mu}$ the renormalized chemical potential. Find the condition that the coefficients need to satisfy for the β-function not to explicitly depend on Λ.

Solution The β-function is

$$\beta(\lambda_r) = \frac{d\lambda_r}{d\ln\left(\frac{\Lambda}{|\tilde{\mu}|^{1/2}}\right)} = a\lambda^2 + \lambda^3\left(2b\ln\left(\frac{\Lambda}{|\tilde{\mu}|^{1/2}}\right) + c\right) + O(\lambda^4)$$

$$= a\lambda_r^2 + \left(2(b - a^2)\ln\left(\frac{\Lambda}{|\tilde{\mu}|^{1/2}}\right) + c\right)\lambda_r^3 + O(\lambda_r^4).$$

For the β-function not to explicitly depend on the ultraviolet cutoff, one needs $b = a^2$. In that case

$$\beta(\lambda_r) = a\lambda_r^2 + c\lambda_r^3 + O(\lambda_r^4).$$

Note how the coefficient of the first subleading term c enters the next-order coefficient in the β-function.

3.7 Computation of anomalous dimension*

The field-theoretic method may be utilized to compute the lowest-order correction to the anomalous dimension η, which, as we saw, is $O(\epsilon^2)$. Consider the two second-order diagrams in Fig. 3.2 (c) and (d): unlike the latter, the former is independent of the wavevector of the external legs, and thus contributes only to $\chi^{-1}(0)$. Only the "sunrise" diagram in Fig. 3.2 (d) therefore contributes to the anomalous dimension. At the transition point $\tilde{\mu} = 0$, to the second order in λ the inverse susceptibility may be written as

$$\chi^{-1}(k) = k^2 - 8(M+1)\frac{\lambda^2}{2}[I(k) - I(0)], \tag{3.73}$$

where

$$I(k) = \int_0^\Lambda \frac{d\vec{p}\,d\vec{q}}{(2\pi)^{2d}} \frac{1}{p^2 q^2 (\vec{k} + \vec{p} - \vec{q})^2}. \tag{3.74}$$

For generality we assumed M complex components, which yields the combinatorial factor of $8(M+1)$ (Problem 3.2). The anomalous dimension is then

$$2 - \eta = \lim_{k \to 0} \frac{d \ln \chi^{-1}(k)}{d \ln k}. \tag{3.75}$$

The principal technical problem in the evaluation of η lies in the computation of the integral $I(k)$, the integrand of which depends not only on the magnitudes of the internal momenta $\vec{p}$ and $\vec{q}$, but on their relative orientation as well. The calculation becomes particularly difficult if $\vec{p}$ and $\vec{q}$ are to be restricted to the momentum shell, which is why it is typically performed using the field-theoretic approach in which the integration is unrestricted.

To evaluate $I(k)$ consider first the simpler integral

$$I_1(k) = \int_0^\Lambda \frac{d\vec{q}}{(2\pi)^d} \frac{1}{q^2 (\vec{k} - \vec{q})^2}, \tag{3.76}$$

which appears as part of it. Since for $d < 4$, $I_1(k)$ is convergent when the upper limit of integration is infinite, we may set $\Lambda = \infty$. It is convenient to use the identity

$$\frac{1}{a^u b^w} = \frac{\Gamma(u+w)}{\Gamma(u)\Gamma(w)} \int_0^1 d\alpha \frac{\alpha^{u-1}(1-\alpha)^{w-1}}{[a\alpha + b(1-\alpha)]^{u+w}}, \tag{3.77}$$

with u and w real and positive and $\Gamma(z)$ as the standard factorial function, to write

$$I_1(k) = \int \frac{d\vec{q}}{(2\pi)^d} \int_0^1 \frac{d\alpha}{[(1-\alpha)q^2 + \alpha(k^2 - 2\vec{k}\cdot\vec{q} + q^2)]^2}. \tag{3.78}$$

The identity in Eq. (3.77) was introduced by R. Feynman to simplify calculations in quantum electrodynamics. Its great advantage is that after exchanging the order of integrals over the wavevector $\vec{q}$ and *Feynman's parameter* α, so that

$$I_1(k) = \int_0^1 d\alpha \int \frac{d\vec{q}}{(2\pi)^d} \frac{1}{[(\vec{q} - \alpha\vec{k})^2 + \alpha(1-\alpha)k^2]^2}, \tag{3.79}$$

the integral over $\vec{q}$ can be straightforwardly performed. Since the range of integration over $\vec{q}$ is now completely unrestricted we are free to change the variables as $\vec{q} - \alpha\vec{k} \to \vec{q}$, and then integrate to obtain

$$I_1(k) = \frac{\Gamma(2 - \frac{d}{2})}{(4\pi)^{d/2}} \int_0^1 d\alpha [\alpha(1-\alpha)k^2]^{\frac{d}{2}-2}. \tag{3.80}$$

Final integration over Feynman's parameter α gives

$$I_1(k) = \frac{\Gamma(2 - \frac{d}{2})\Gamma^2(\frac{d}{2} - 1)}{(4\pi)^{d/2}\Gamma(d - 2)} k^{-\epsilon}, \tag{3.81}$$

with $\epsilon = 4 - d$, as usual. The inverse susceptibility is therefore

$$\chi^{-1}(k) = k^2 - 4(M+1)\lambda^2 \frac{\Gamma(2 - \frac{d}{2})\Gamma^2(\frac{d}{2} - 1)}{(4\pi)^{d/2}\Gamma(d - 2)}(I_2(k) - I_2(0)), \tag{3.82}$$

where

$$I_2(k) = \int \frac{d\vec{p}}{(2\pi)^d} \frac{1}{p^2[(\vec{k} + \vec{p})^2]^{\frac{\epsilon}{2}}}. \tag{3.83}$$

To compute $I_2(k)$ we use Feynman's identity once again and write

$$I_2(k) = \frac{\Gamma(1 + \frac{\epsilon}{2})}{\Gamma(\frac{\epsilon}{2})} \int_0^1 d\alpha \, \alpha^{\frac{\epsilon}{2}-1} \int_0^\Lambda \frac{d\vec{p}}{(2\pi)^d} \frac{1}{[(\vec{p} + \alpha\vec{k})^2 + \alpha(1-\alpha)k^2]^{1+\frac{\epsilon}{2}}}. \tag{3.84}$$

If we took $\Lambda = \infty$ again, as we did in the evaluation of $I_1(k)$, the integral is convergent only in dimensions $d < 3$. The same ultraviolet divergence for $d \geq 3$, however, is present in $I_2(0)$ as well, and so cancels out in the difference $I_2(k) - I_2(0)$ that we actually need. The finite residue is most elegantly obtained by using the formalism of *dimensional regularization*. For $z > d/2$ we have

$$\int \frac{d\vec{p}}{(2\pi)^d} \frac{1}{(p^2 + m^2)^z} = \frac{\Gamma(z - \frac{d}{2})m^{d-2z}}{(4\pi)^{d/2}\Gamma(z)}, \tag{3.85}$$

where the region of integration in the integral in the last equation is completely unrestricted. For $z < d/2$, on the other hand, the integral is evidently divergent. For $z < d/2$ we will then use the analytic continuation of the expression on the right hand side to *define* what is meant by the diverging integral on the left hand side. Note that according to this definition the integral vanishes when $m = 0$ for $z < d/2$. So by the rules of dimensional regularization the infinite $I_2(0)$ is formally set to zero, to be automatically left only with the finite difference $I_2(k) - I_2(0)$ in applying it to $I_2(k)$.

Assuming dimensional regularization, we then set the upper limit on integration in $I_2(k)$ in Eq. (3.84) to infinity, and obtain

$$I_2(k) = \frac{\Gamma(3-d)\Gamma(\frac{d}{2}-1)\Gamma(d-2)}{\Gamma(2-\frac{d}{2})\Gamma(3(\frac{d}{2}-1))(4\pi)^{d/2}} k^{2-2\epsilon}. \tag{3.86}$$

Inserting this back into Eq. (3.82), and computing the anomalous dimension from Eq. (3.75), then yields

$$\eta = 4(M+1)\frac{\Gamma^3(\frac{d}{2}-1)\Gamma(3-d)}{\Gamma(3(\frac{d}{2}-1))} \lim_{k\to 0} \frac{d\hat{\lambda}^2}{d\ln k} + O(\hat{\lambda}^3), \tag{3.87}$$

with dimensionless $\hat{\lambda} = \lambda k^{-\epsilon}/(4\pi)^{d/2}$. Using the Laurent expansion of the factorial function,

$$\Gamma(-n+x) = \frac{(-1)^n}{n!}\left(\frac{1}{x} + \Psi(n+1) + O(x)\right), \tag{3.88}$$

where $n \geq 0$ is an integer and $\Psi(n+1) = 1 + (1/2) + \cdots + (1/n) - 0.577$, we find

$$\eta = 4(M+1)\lim_{k\to 0}\hat{\lambda}^2 \tag{3.89}$$

to the lowest order in $\hat{\lambda}$ and ϵ. Similarly as for the exponent γ in Eq. (3.61), the perturbative result for η is infinite as well. We expect that the series in $\hat{\lambda}$ can be turned into one for the renormalized interaction $\hat{\lambda}_r$, which will have a finite infrared limit when obtained from the β-function. One cannot, however, readily use the β-function in Eq. (3.68), since, being right at the transition point $\tilde{\mu} = 0$, the current definition of the dimensionless interaction $\hat{\lambda}$ differs from the one employed in the computation of γ. We therefore need first to define the renormalized interaction at the transition point: take λ_r to be the value of the interaction vertex at $\tilde{\mu} = 0$ and for the finite value of the external wavevectors:

$$\vec{k}_i \cdot \vec{k}_j = \left(\delta_{ij} - \frac{1}{4} \right) k^2, \tag{3.90}$$

for $i, j = 1, 2, 3$, and $\vec{k}_1 + \vec{k}_2 = \vec{k}_3 + \vec{k}_4$. This point in wavevector space is for obvious reasons called the *symmetric point*. The choice of the symmetric point is arbitrary, and made only to simplify the calculation. Reevaluating the diagrams in Fig. 3.2 (a) and (b) at the symmetric point and for $\tilde{\mu} = 0$ leads to

$$\lambda_r = \lambda - \left(\frac{M+3}{2^{2-\frac{d}{2}}} + 1 \right) \frac{2\Gamma(2-\frac{d}{2})\Gamma^2(\frac{d}{2}-1)}{\Gamma(d-2)} \frac{\lambda^2}{(4\pi)^{d/2}k^\epsilon}. \tag{3.91}$$

Defining $\hat{\lambda}_r = \lambda_r k^{-\epsilon}/(4\pi)^{d/2}$, we find

$$\hat{\lambda}_r = \hat{\lambda} - 4(M+4) \left(\frac{1}{\epsilon} + O(1) \right) \hat{\lambda}^2 + O(\hat{\lambda}^3), \tag{3.92}$$

and thus

$$\beta(\hat{\lambda}_r) = \epsilon\hat{\lambda}_r - 4(M+4)\hat{\lambda}_r^2 + O(\hat{\lambda}^3). \tag{3.93}$$

The anomalous dimension is then

$$\eta = 4(M+1) \lim_{k\to 0} \hat{\lambda}_r^2 = \frac{M+1}{4(M+4)^2}\epsilon^2 + O(\epsilon^3). \tag{3.94}$$

The result for N real components may be obtained by replacing M in the above equation by $N/2$. Note also that after all the care exercised in defining the renormalized interaction properly at the transition point, the β-function in Eq. (3.93), upon accounting for the factor of two difference in definitions of the coupling constants as $\epsilon \to 0$, is in fact identical to the the one in Eq. (3.68).

Problem 3.6* Show that the first two coefficients in the Taylor series for the β-function at the upper critical dimension are invariant under finite redefinitions of the coupling constant.

Solution Assume that

$$\beta(\lambda) = a\lambda^2 + b\lambda^3 + O(\lambda^4)$$

and define a new coupling constant g as

$$\lambda = g + xg^2 + O(g^3),$$

where x is a finite number. Then

$$\beta(\lambda) = (1 + 2xg + O(g^2))\beta(g)$$

and therefore

$$\beta(g) = (1 - 2xg + O(g^2))(ag^2 + (2xa + b)g^3 + O(g^4))$$
$$= ag^2 + bg^3 + O(g^4).$$

3.8 Summary

Near and below upper critical dimension the effect of the elimination of high-energy degrees of freedom in the Ginzburg–Landau–Wilson theory is indeed only the renormalization of the already existing couplings in the action. In particular, all couplings absent initially, although generated in the process of mode elimination, are ultimately irrelevant. This explains the scaling forms of the physical quantities like the free energy and the susceptibility near the transition. The only relevant couplings at the critical point are the chemical potential (i.e. temperature) and the source (or magnetic field), while the deviation of the quartic interaction from its fixed point value is the least irrelevant variable, and as such provides the leading correction to scaling. Above the upper critical dimension, on the other hand, a weak interaction is irrelevant and the critical behavior is of the mean-field type. However, the interaction is dangerously irrelevant for $d > 4$, which leads to violation of hyperscaling.

What happens when the results are extended to physical dimensions $d = 3$ or $d = 2$? For N real components one finds

$$\nu = \frac{1}{2} + \frac{N+2}{4(N+8)}\epsilon + \frac{(N+2)(N^2+23N+60)}{8(N+8)^3}\epsilon^2 + O(\epsilon^3), \quad (3.95)$$

$$\gamma = 1 + \frac{N+2}{2(N+8)}\epsilon + \frac{(N+2)(N^2+22N+52)}{4(N+8)^3}\epsilon^2 + O(\epsilon^3). \quad (3.96)$$

The series have actually been computed to fifth order in ϵ. The case of the Ising model ($N = 1$) in $d = 3$, for example, gives the following sequence of values for γ: 1, 1.167, 1.244, 1.195, 1.375, 0.96 from the zeroth to the fifth order in ϵ, respectively. The almost exact value $\gamma = 1.239 \pm 0.003$ for the Ising model in $d = 3$ comes from the high-temperature expansion. When the

Table 3.1 *The correlation length exponent obtained from the*
Borel-summed ϵ-expansion (ϵ), the Borel-summed field-theoretic
renormalization group in fixed dimension (RG), numerical Monte Carlo
calculations (MC), experiment (EXP), and the high-temperature
expansion (HT), for the Ising ($N = 1$), the superfluid ($N = 2$), and the
Heisenberg ($N = 3$) universality classes.

ν	$N = 1$	$N = 2$	$N = 3$
ϵ	0.6290 ± 0.0025	0.6680 ± 0.0035	0.7045 ± 0.0055
RG	0.6304 ± 0.0013	0.6703 ± 0.0015	0.7073 ± 0.0035
MC	0.628 ± 0.004	0.665 ± 0.008	0.707 ± 0.006
EXP	0.64 ± 0.01	0.6706 ± 0.0006	0.69 ± 0.02
HT	0.631 ± 0.004	0.670 ± 0.007	0.72 ± 0.01

ϵ-expansion is compared with this result one finds that, after coming very close at the second-order, the series starts deviating significantly without any signs of convergence. In fact, the ϵ-expansion is an example of a divergent series, but it is believed to be asymptotic and Borel-summable. For an asymptotic series the error at any given order is bounded by the next-order term, so one should keep adding terms while they are decreasing. After that point, however, adding further terms makes the result of the series less accurate. In the above series for γ this presumably happens at the second order.

One can nevertheless extract very accurate results from the above series using Borel summation, in which one makes an educated guess about the function being expanded from the finite number of computed terms. In Table 3.1 we provide a summary of the results for the correlation length exponent in $d = 3$ from various methods.[2] The overall agreement must be judged as excellent. Likewise, even in $d = 2$ the Borel-summed ϵ-expansion leads to $\gamma = 1.73 \pm 0.06$ and $\eta = 0.26 \pm 0.05$, for $N = 1$, in very good agreement with the exact solution of the Ising model which yields $\gamma = 1.75$ and $\eta = 0.25$.

The main strength of the momentum-shell or the field-theoretic renormalization group and the concomitant ϵ-expansion lies not in its accuracy in computing the critical exponents. Indeed, there are more efficient methods for that, like the renormalization group in fixed dimension, which is

---

[2] Extracted from the summary by R. Guida and J. Zinn-Justin, *Journal of Physics A: Mathematical and General* **31**, 8103 (1998).

easier to handle at higher orders and may also be proven to be Borel-summable. The true value of the ϵ-expansion lies in its controlled nature, which allows one to understand the flow diagram and classify the couplings according to their relevancy. One often relies on the structure obtained in the ϵ-expansion when applying other methods in physical dimensions. There are examples, however, when the low-order ϵ-expansion is even qualitatively wrong when naively extended to physical dimensions. An important example of this is provided by the superconducting transition discussed in the next chapter.

Problem 3.7 Show that the slope of $\beta_\lambda = d\lambda/d\ln(b)$ at the critical point is invariant under a redefinition of the coupling constant $\lambda \to \lambda' = f(\lambda)$.

Solution Since

$$\beta_{\lambda'} = \frac{d\lambda'}{d\ln b} = \frac{d\lambda'}{d\lambda}\beta_\lambda,$$

it follows that

$$\frac{d\beta_{\lambda'}}{d\lambda'} = \frac{d^2\lambda'}{d\lambda^2}\frac{\beta_\lambda}{d\lambda'/d\lambda} + \frac{d\beta_\lambda}{d\lambda}.$$

At $\lambda'^* = f(\lambda^*)$ the first term vanishes and the two slopes, which determine the correction to scaling exponent, become equal. This is the simplest example of invariance of critical exponents under a redefinition of coupling constants.

Problem 3.8 Find the critical exponents in $d = 4 - \epsilon$ dimensions in the $U(N) \times U(N)$-symmetric Ginzburg–Landau–Wilson theory with the action

$$S = \int d\vec{r}\left[\sum_{i=1}^{2}\left(|\nabla\Phi_i|^2 - \mu|\Phi_i|^2 + \frac{\lambda_2}{2}|\Phi_i|^4\right) + \frac{\lambda_1}{2}\left(\sum_{i=1}^{2}|\Phi_i|^2\right)^2\right],$$

where $\Phi_i(\vec{r})\ i = 1, 2$ has N complex components.

Solution It is convenient to define a new coupling $\lambda = (\lambda_1 + \lambda_2)/2$, so that the quartic term in the action becomes $\lambda(|\Phi_1|^4 + |\Phi_2|^4) + \lambda_1|\Phi_1|^2|\Phi_2|^2$. These terms then may be represented diagrammatically by the three different vertices shown in Fig. 3.8.

Besides the diagrams in Fig. 3.2 (a) and (b) for both quartic vertices, there are six additional diagrams depicted in Fig. 3.9. These yield the contributions

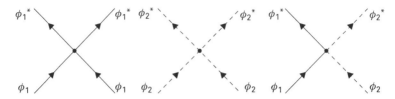

Figure 3.8 Three interaction vertices in the $U(N) \times U(N)$ action S.

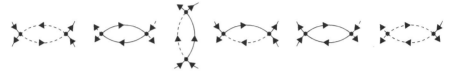

Figure 3.9 Lowest-order contributions to the renormalized interactions in the $U(N) \times U(N)$ Ginzburg–Landau–Wilson theory.

$-N\lambda_1^2 \ln(b)/2$ to $\lambda(b)$ ((a) and (b)), $-2\lambda_1^2 \ln(b)$ ((c) and (d)) and $-4\lambda\lambda_1(N + 1)\ln(b)$ ((e) and (f)) to $\lambda_1(b)$. Here, as usual, the factor $\Lambda^{d-4}S_d/(2\pi)^d$ has been absorbed into the coupling constants.

So,

$$\frac{\mathrm{d}\lambda}{\mathrm{d}\ln(b)} = \epsilon\lambda - 2(N+4)\lambda^2 - \frac{N}{2}\lambda_1^2,$$

$$\frac{\mathrm{d}\lambda_1}{\mathrm{d}\ln(b)} = \epsilon\lambda_1 - 2\lambda_1^2 - 4(N+1)\lambda_1\lambda,$$

or, in terms of the original couplings λ_1 and λ_2,

$$\frac{\mathrm{d}\lambda_1}{\mathrm{d}\ln(b)} = \epsilon\lambda_1 - 2(N+2)\lambda_1^2 - 2(N+1)\lambda_1\lambda_2,$$

$$\frac{\mathrm{d}\lambda_2}{\mathrm{d}\ln(b)} = \epsilon\lambda_2 - (N+4)\lambda_2^2 - 6\lambda_1\lambda_2.$$

There exist therefore four fixed points: (1) Gaussian, at $\lambda_1 = \lambda_2 = 0$, (2) $U(2N)$ symmetric, at $\lambda_2 = 0$, $\lambda_1 = \epsilon/(2(N+2))$, (3) decoupled fixed point, at $\lambda_2 = \epsilon/(N+4)$ and $\lambda_1 = 0$, and (4) mixed, at $\lambda_1 = \epsilon(2 - N)/(2(N^2 + 2))$ and $\lambda_2 = \epsilon(N - 1)/(N^2 + 2)$. The flows for different N are depicted in Fig. 3.10. For $N < 1$, the stable fixed point is the one with the enlarged $U(2N)$ symmetry. At this fixed point the correlation length exponent is given by the result in Problem 3.1, with M replaced by $2N$. For $1 < N < 2$, on the other hand, the mixed fixed point becomes the stable one. Finally, for $N > 2$ the decoupled fixed point is stable.

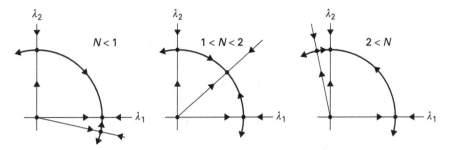

Figure 3.10 The flow of the interaction coupling constants in the critical plane $\tilde{\mu} = 0$ for different numbers of component in the $U(N) \times U(N)$ Ginzburg–Landau–Wilson theory.

The flow of the chemical potential to the first order in interactions is in general given by

$$\frac{d\tilde{\mu}}{d\ln(b)} = \tilde{\mu}(2 - (2N+1)\lambda_1 - (N+1)\lambda_2),$$

and therefore the correlation length exponent at the mixed fixed point is

$$\nu = \frac{1}{2} + \frac{3N}{8(N^2+2)}\epsilon + O(\epsilon^2).$$

The anomalous dimension is $\eta = O(\epsilon^2)$ at all N. The remaining exponents may be found from the scaling relations.

Problem 3.9 Determine the leading correction to the scaling exponent at the mixed fixed point in the previous problem.

Solution The stability matrix has the elements $M_{11} = 1/\nu$, $M_{12} = M_{13} = M_{21} = M_{31} = 0$, and $M_{22} = \epsilon(N^2 - 4)/(N^2 + 2)$, $M_{23} = \epsilon(N - 2)(N + 1)/(N^2 + 2)$, $M_{32} = 6\epsilon(1 - N)/(N^2 + 2)$, and $M_{33} = -\epsilon(N^2 + 3N - 4)/(N^2 + 2)$. Besides M_{11}, the other two eigenvalues are then $-\epsilon$ and $(N^2 - 3N + 2)\epsilon/(N^2 + 2)$. While the first one is always negative, the second is negative only for $1 < N < 2$, when the mixed fixed point is critical. The correction to the scaling exponent for $1 < N < 2$ is therefore

$$\omega = \frac{-2 + 3N - N^2}{2(N^2 + 2)}\epsilon + O(\epsilon^2).$$

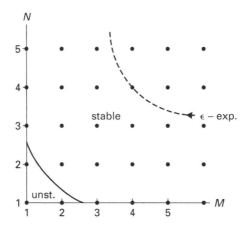

Figure 3.11 The region of stability of the decoupled fixed point in the $O(N) \times O(M)$ model from the exact criterion (full line), and the ϵ-expansion (dotted line).

Problem 3.10* Note that the β-function for λ_1 at the decoupled fixed point in Problem 3.8 to the lowest order in ϵ may be written as

$$\frac{d\lambda_1}{d\ln(b)} = \frac{\alpha}{\nu}\lambda_1 + O(\lambda_1^2),$$

where α and ν are the critical exponents in the $U(N)$ Ginzburg–Landau–Wilson theory. Generalize this result and use it to determine the stability of the decoupled fixed point at $u = u^*$, $v = v^*$, $w^* = 0$ in the three dimensional $O(N) \times O(M)$ Ginzburg–Landau–Wilson theory

$$S = \int d\vec{r}[(\nabla\Phi_1)^2 + (\nabla\Phi_2)^2 + a_1(\Phi_1)^2 + a_2(\Phi_2)^2 + u(\Phi_1)^4 + v(\Phi_2)^4$$
$$+ 2w(\Phi_1)^2(\Phi_2)^2],$$

where $\Phi_1(\Phi_2)$ has $N(M)$ real components.

Solution At the decoupled fixed point we have $(\Phi_1)^2 \sim b^{(1/\nu_N)-d}$, and analogously for Φ_2. Since $wb^d(\Phi_1)^2(\Phi_2)^2 \sim 1$, using the hyperscaling relation we find that the above result generalizes into

$$\frac{dw}{d\ln(b)} = \frac{1}{2}\left(\frac{\alpha_N}{\nu_N} + \frac{\alpha_M}{\nu_M}\right)w + O(w^2),$$

where α_N and ν_N are the critical exponents at the $O(N)$ critical point. This relation is satisfied order by order in the ϵ-expansion to the order ϵ^5, and is believed to be exact. Using the values for ν in Table 3.1 and the hyperscaling relation, we find the decoupled fixed point to be unstable only for $N = M = 1$, and $N = 1$, $M = 2$, and stable otherwise. The region of stability of the decoupled fixed point in the $N - M$ plane is sketched in Fig. 3.11. The lowest order ϵ-expansion result, which severely underestimates the region of stability, is also given for comparison.

For $N = M = 1$ high-order calculations show that the stable fixed point is the $O(N + M)$ symmetric, while for $N = 1$ and $M = 2$ the stable fixed point is the mixed one. Although the exact proof is still lacking, experience suggests that there is always exactly one stable fixed point in the theory.

4

Superconducting transition

The Ginzburg–Landau theory describing the Meissner transition in super-conductors is introduced, and two types of superconductor are defined. It is shown that fluctuations of the gauge field lead to first-order transition in type-I superconductors. Calculation near four dimensions is performed for type-II superconductors, and the dependence of the flow diagram on the number of components is discussed. Scaling of the correlation length and of the penetration depth near the transition is elaborated.

4.1 Meissner effect

Most elemental metals and many alloys go through a sharp phase transition in which the material becomes a perfect diamagnet at low magnetic fields and completely loses its electrical resistance when cooled down to temperatures of several kelvins (Fig. 4.1). Such a "superconducting" transition has now been observed at temperatures as high as ~ 150 K, in materials known as high-temperature superconductors. Superconductivity is a closely related phenomenon to superfluidity in ^{4}He, except that electrons are charged and as such carry electrical current. Even before the advance of the microscopic theory of superconductivity in metals and alloys, V. Ginzburg and L. Landau devised a phenomenological description of the transition and the superconducting state. Cast in modern language, their proposal is that the finite temperature superconducting transition may be described by the partition function

$$Z = \int D\Phi^* D\Phi D\vec{A} \exp\left[-\left(\int d\vec{x} L_{GL}[\Phi, \vec{A}]\right)/T\right], \qquad (4.1)$$

77

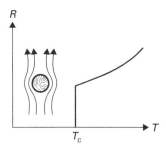

Figure 4.1 Resistivity as a function of temperature near superconducting transition. Above critical temperature the material is a metal. Below T_c the low magnetic field is expelled from the interior of the superconductor (the "Meissner effect").

with

$$L_{\text{GL}} = |(\nabla - ie\vec{A}(\vec{x}))\Phi(\vec{x})|^2 + a(T)|\Phi(\vec{x})|^2 + \frac{b}{2}|\Phi(\vec{x})|^4 + \frac{1}{2}(\nabla \times \vec{A}(\vec{x}))^2,$$

(4.2)

where $\Phi(\vec{x})$ is a complex function of spatial coordinates, and $\vec{A}(\vec{x})$ is the fluctuating electromagnetic vector potential. Here the coefficients $a(T) = a_0(T - T_0)$, a_0 and b are constants, and e is *twice* the electron charge. We have also set $\hbar = c = k_B = 1$ for simplicity, and as usual assumed an ultraviolet cutoff Λ on the wavevectors. Once the microscopic theory of superconductivity was developed it was shown that the Ginzburg–Landau theory may indeed be derived from it, and the coefficients in L_{GL} calculated in terms of the microscopic parameters of the system. A closely related partition function describes also the nematic-to-smectic-A transition in liquid crystals.

The Ginzburg–Landau theory may be the simplest representative of the class of so-called *gauge theories*, which are believed to describe all the fundamental interactions of matter at the energies accessible to today's accelerators. In a somewhat more elaborate form it appears as an essential ingredient of the "standard model" of elementary particles, where it is usually called the *Higgs model*. From the point of view of critical phenomena it represents a particularly interesting example of a phase transition which displays several non-trivial features in a relatively simple context, which makes it worth understanding in some detail.

For $e = 0$, L_{GL} becomes equivalent to Landau's action introduced in Chapter 2 which described the transition in an electrically neutral superfluid, such as ^{4}He. The main novelty is that superconductors are charged, so that the complex order parameter is coupled to fluctuations in the electromagnetic vector potential. This leads to the *Meissner effect*, that is to the expulsion of a weak external magnetic field from the interior of the superconductor, which is the defining feature of the superconducting state.

Let us assume $T < T_0$ and expand the action around the saddle-point value of the field $\Phi_0 = \sqrt{|a(T)|/b}$, as

$$\Phi(\vec{x}) = e^{i\phi(\vec{x})}(\Phi_0 + \psi(\vec{x})). \tag{4.3}$$

First, let us consider the neutral superfluid, and $e = 0$. The result of the expansion is

$$L_{GL} = L_0 + |\Phi_0|^2(\nabla\phi(\vec{x}))^2 + (\nabla\psi(\vec{x}))^2 + 2|a(T)|\psi(\vec{x})^2$$
$$+ \frac{1}{2}(\nabla \times \vec{A}(\vec{x}))^2 + O(\psi^4, \phi^4, \phi^2\psi^2). \tag{4.4}$$

We may say that the amplitude fluctuations ψ in the ordered state are "massive": they cost a finite energy $2|a(T)|$ even when they vary arbitrarily slowly in space. The phase fluctuations ϕ, on the other hand, are "massless": their energy vanishes when $\nabla\phi \rightarrow 0$. This is the consequence of the global $U(1)$ symmetry of Landau's ($e = 0$) action; rotating uniformly $\Phi_0 \rightarrow e^{i\phi}\Phi_0$ with ϕ constant in space cannot cost any energy. This is an example of *Nambu–Goldstone's theorem*: whenever a continuous global symmetry is spontaneously broken there exist massless modes in the ordered state. Note also that when $e = 0$ the vector potential $\vec{A}$ is massless both above and below the transition.

A more detailed discussion of Goldstone's modes will be presented in the next chapter. Here we are interested in the fate of the massless phase mode when $e \neq 0$. First, notice that for $e \neq 0$ the action becomes symmetric under a much larger *local* (space-dependent) $U(1)$ transformation

$$\Phi(\vec{x}) \rightarrow e^{i\phi(\vec{x})}\Phi(\vec{x}), \tag{4.5}$$

provided it is accompanied by the *gauge transformation*

$$\vec{A}(\vec{x}) \rightarrow \vec{A}(\vec{x}) + \frac{1}{e}\nabla\phi(\vec{x}). \tag{4.6}$$

After the expansion of $\Phi(\vec{x})$ around Φ_0 as in Eq. (4.3), the phase $\phi(\vec{x})$ may therefore be completely absorbed into the redefined vector potential. So, when $e \neq 0$,

$$L = L_0 + (\nabla\psi(\vec{x}))^2 + 2|a(T)|\psi^2(\vec{x}) + O(\psi^4)$$
$$+ e^2|\Phi_0|^2\vec{A}^2(\vec{x}) + \frac{1}{2}(\nabla \times \vec{A}(\vec{x}))^2. \tag{4.7}$$

In contrast to the neutral ($e = 0$) case, there are no massless modes in the superconducting phase: ϕ has completely disappeared from L. However, the electromagnetic vector potential $\vec{A}$ has become massive. This is the simplest example of the *Higgs mechanism*, believed today to be the source of mass of the particles mediating weak force in the standard model of elementary particles. In the superconducting phase the correlation function of the magnetic field from Eq. (4.7) is then

$$\langle (\nabla \times \vec{A}(\vec{x}))_i (\nabla \times \vec{A}(0))_j \rangle = \int d^3 q \frac{q^2(\delta_{ij} - \hat{q}_i \hat{q}_j)}{q^2 + 2e^2|\Phi_0|^2} e^{i\vec{q}\cdot\vec{x}} \sim \frac{\hat{x}_i \hat{x}_j - \delta_{ij}}{4\pi \lambda^2 |\vec{x}|} e^{-x/\lambda},$$

(4.8)

for $|\vec{x}|/\lambda \gg 1$, where $\lambda = 1/\sqrt{2e^2|\Phi_0|^2}$ is the magnetic field *penetration depth*. The correlations of the magnetic field decay exponentially over the length λ. This implies that a weak external magnetic field cannot penetrate the interior of the superconductor of a size much larger than the penetration depth, i.e. perfect diamagnetism. This is the Meissner effect.

Superconductors are therefore characterized by *two* characteristic temperature-dependent lengths: the correlation length $\xi^{-2} = |a(T)|$ that describes the spatial dependence of the order parameter fluctuations, and the penetration depth $\lambda^{-2} = 2e^2|a(T)|/b$, that enters the magnetic field spatial fluctuations. Their dimensionless ratio

$$\kappa = \frac{\lambda}{\xi} = \sqrt{\frac{b}{2e^2}}$$

(4.9)

defines an important characteristic of a superconductor, called the Ginzburg–Landau parameter. The response of the superconductor to an external magnetic field is well known to depend crucially on the value of κ: for $\kappa < 1/\sqrt{2}$ the transition from the Meissner phase into the normal phase with increase in the magnetic field is direct and discontinuous, whereas for $\kappa > 1/\sqrt{2}$ there is a superconductor–normal mixed phase in between. The two different classes of materials are called type-I and type-II; the former includes all the metals, while the latter includes many alloys and the high-temperature superconductors. The discussion of superconductivity in a magnetic field may be found in many books on solid state physics, and will not be pursued further here. In the following we discuss the role the Ginzburg–Landau parameter plays at the superconducting transition even at zero external magnetic field.

Problem 4.1 Compute the magnetic field correlation function in the super-conducting phase in Eq. (4.8) at an arbitrary distance $\vec{x}$.

Solution The Fourier integral in Eq. (4.8) can be transformed into

$$\int d^3\vec{q}\, \frac{q^2(\delta_{ij} - \hat{q}_i\hat{q}_j)}{q^2 + 2e^2|\Phi_0|^2} e^{i\vec{q}\cdot\vec{x}} = \delta_{ij}\delta(\vec{x}) + \left(\frac{\partial^2}{\partial x_i \partial x_j} - \frac{\delta_{ij}}{\lambda^2}\right)\frac{e^{-x/\lambda}}{4\pi x}.$$

Taking the derivatives yields

$$\langle\langle (\nabla \times \vec{A}(\vec{x}))_i (\nabla \times \vec{A}(0))_j \rangle\rangle = \delta_{ij}\delta(\vec{x}) + \left[(3\hat{x}_i\hat{x}_j - \delta_{ij})(1 + (x/\lambda))\right.$$
$$\left. + (\hat{x}_i\hat{x}_j - \delta_{ij})(x/\lambda)^2\right]\frac{e^{-x/\lambda}}{4\pi x^3}.$$

For $x/\lambda \gg 1$ this reduces to the result quoted in the text.

4.2 Fluctuation-induced first-order transition

The nature of the superconducting transition is a rather subtle issue, and has been a problem in the theory of critical phenomena for some time. The subtlety arises due to the coupling of the superconducting order parameter to the fluctuating gauge-field, which is massless in the normal phase. Issues similar to the ones that will be discussed in the rest of this chapter may thus be expected to arise whenever there is such a coupling of the order parameter to massless modes. This type of situation, for example, arises generically in quantum phase transitions of interacting itinerant electrons, where the order parameter is coupled to particle–hole excitations of the electronic Fermi sea.

Since L_{GL} is quadratic in the vector potential one may try to integrate it out. This can be done in the mean-field approximation for the field Φ, in which one assumes a spatially constant value of Φ, integrates over the vector potential, and then determines Φ at the saddle point by minimizing the resulting action. The integration over the vector potential will be done in the *transverse gauge* $\nabla \cdot \vec{A} = 0$, in which

$$\langle A_i(\vec{q})A_j(-\vec{q})\rangle = (2\pi)^d \frac{\delta_{ij} - \hat{q}_i\hat{q}_j}{q^2 + 2e^2|\Phi|^2}. \tag{4.10}$$

The choice of transverse gauge in the present example is more than just a matter of convenience: as discussed in Appendix C, it is only in this gauge that the usual correlation function $\langle\Phi^*(\vec{x})\Phi(0)\rangle$ becomes long ranged below T_c. The resulting free energy per unit volume for a uniform Φ and in $d = 3$ after the integration over the two transverse components of the vector potential is

$$F = a(T)|\Phi|^2 + \frac{b}{2}|\Phi|^4 + T_0 \int \frac{d^3\vec{q}}{(2\pi)^3} \ln(q^2 + 2e^2|\Phi|^2). \tag{4.11}$$

Figure 4.2 Free energy of the superconductor after the integration over the gauge field, above, at, and below the critical temperature. The superconducting transition at $T = T_c$ is discontinuous.

Taking the derivative and rearranging the terms slightly, we find

$$\frac{\mathrm{d}F}{\mathrm{d}|\Phi|} = 2\left(a(T) + 2T_0 e^2 \int \frac{\mathrm{d}^3\vec{q}}{(2\pi)^3} \frac{1}{q^2}\right)|\Phi| + 2b|\Phi|^3$$

$$- 8e^4 T_0 \int \frac{\mathrm{d}^3\vec{q}}{(2\pi)^3} \frac{|\Phi|^3}{q^2(q^2 + 2e^2|\Phi|^2)}. \tag{4.12}$$

The integral in the first term therefore simply lowers the transition temperature by a finite amount. We will assume this shift to be included in the definition of T_0 hereafter. Performing the integral in the last term and assuming $\Lambda/(e\Phi) \gg 1$ near the transition,

$$\frac{\mathrm{d}F}{\mathrm{d}|\Phi|} = 2a(T)|\Phi| + 2b|\Phi|^3 - \frac{\sqrt{2}e^3 T_0}{\pi}|\Phi|^2, \tag{4.13}$$

and finally, integrating,

$$F = a(T)|\Phi|^2 - \frac{\sqrt{2}e^3 T_0}{3\pi}|\Phi|^3 + \frac{b}{2}|\Phi|^4. \tag{4.14}$$

The mean-field free energy now contains a negative term cubic in $|\Phi|$. This term is evidently non-analytic in $|\Phi|^2$, which is possible only because the electromagnetic potential that was integrated out was massless in the normal phase. The value of Φ is determined by the equation $\mathrm{d}F/\mathrm{d}\Phi = 0$. At high temperatures the only solution is $\Phi = 0$. With decrease in temperature two new solutions emerge, the first describing a local maximum, and the second a local minimum of the action (Figure 4.2). Below the critical temperature the free energy at the non-trivial minimum becomes negative, and the system suffers a *discontinuous* transition from the normal state. The critical temperature T_c and the value of $\Phi \neq 0$ at T_c are determined by the equations $F = 0$

and $dF/d\Phi = 0$. Eliminating Φ gives

$$a(T_c) = \frac{e^6 T_0^2}{(3\pi)^2 b}. \tag{4.15}$$

To check the consistency of the assumption that fluctuations in Φ may be neglected during the integration over the electromagnetic potential, it is useful to determine the *size* of the first-order transition, which we define as $t_{fo} = (T_c - T_0)/T_0$. From Eq. (4.15) it follows that

$$t_{fo} = \frac{b^2 T_0^2}{72\pi^2 |a(0)|} \kappa^{-6}. \tag{4.16}$$

We may recognize $|a(0)| = \xi_0^{-2}$, with ξ_0 being the correlation length at $T = 0$. Recalling that in $d = 3$ the dimensionless quartic interaction may be defined as $\hat{b} = b\xi T_0/(2\pi^2)$ (Eq. 3.62), the above expression may be rewritten as

$$t_{fo} = \frac{\pi^2}{18} \hat{b}_0^2 \kappa^{-6}. \tag{4.17}$$

A calculation to lowest order in ϵ in the previous chapter gave $\hat{b} \approx 1/10$ in the critical region at $e = 0$, where the scaling was governed by the Wilson–Fisher fixed point. The critical region in which fluctuations of the order parameter would become important is therefore

$$t_{cr} \approx 100 \hat{b}_0^2. \tag{4.18}$$

Neglecting the numerical prefactor in Eq. (4.17), we can then write

$$t_{fo} \approx \frac{t_{cr}}{100\kappa^6}. \tag{4.19}$$

The neglect of fluctuations in Φ is thus well justified for good type-I superconductors in which $\kappa < 1$. In this case the first-order transition indeed occurs at temperatures well above the critical region for fluctuations of the order parameter. For type-II superconductors, on the other hand, fluctuations in Φ cannot be neglected, since $t_{fo} < t_{cr}$, and the first-order transition would occur within the critical region.

Before leaving this section it is instructive to express Eq. (4.16) in yet another form. Since in the mean-field free energy per unit volume for $e = 0$ below T_0 is $-a^2(T)/2b$, differentiating twice with respect to temperature yields the jump in specific heat per unit volume in the mean-field approximation at $e = 0$ to be $\Delta C = a_0^2 T_0/b$. So we can write

$$t_{cr} = \frac{100|a(0)|^3}{(2\pi^2)^2(\Delta C)^2} = \frac{100}{4\pi^4(\Delta C\xi_0^3)^2}. \tag{4.20}$$

The quantity $\Delta C \xi_0^3$ is essentially the number of degrees of freedom in the correlation volume participating in the transition. For weakly coupled superconductors like metals, $\xi_0 \sim 10^3 A$, and $t_{cr} \sim 10^{-16}$. This is why elemental superconductors are extremely well described by the mean-field theory at all important temperatures. For superfluids or magnetic systems, on the other hand, $\xi_0 \sim 1A$ and $t_{cr} \sim 1$, and critical fluctuations are readily observable. The relation (4.20) is also known as the *Ginzburg criterion*, and provides a useful estimate of the importance of critical fluctuations at a given phase transition.

By inserting the above estimate of t_{cr} in metals into the Eq. (4.19) we find

$$t_{fo} \approx 10^{-6} \tag{4.21}$$

for $\kappa \approx 0.02$ in aluminum, for example. Since $T_0 \sim 1.19 K$ in aluminum, $T_c - T_0 \sim 1\mu K$, which appears to be unobservably small at the moment. This explains why the fluctuation-induced first-order transition has not yet been seen in superconductors.[1]

Problem 4.2 Find the dependence of the size of the fluctuation-induced first-order transition t_{fo} on the density of electrons.

Solution Since $\kappa = \lambda_0 / \xi_0$ one can rewrite the size of the first-order transition as

$$t_{fo} \sim (\Delta C \lambda_0^3)^{-2},$$

which shows it to be inversely proportional to the number of relevant degrees of freedom in the volume determined by the $T = 0$ penetration depth. Since in weakly coupled superconductors $\Delta C \sim n$ and $\lambda_0^{-2} \sim n$, where n is the electron density, $t_{fo} \sim n$.

Problem 4.3 The application of the external magnetic field in liquid crystals may be represented by a mass term $(H/2)\vec{A}^2$ for the vector field in L_{GL} in Eq. (4.2). Determine the critical value of the magnetic field at which the nematic–smectic-A transition becomes continuous in a good type-I material.

---

[1] In liquid crystals $t_{fo} \sim 10^{-3}$ and the fluctuation-induced first-order nematic-to-smectic-A transition has indeed been observed.

Solution With the mass term for the vector field the free energy in Eq. (4.11) becomes replaced by

$$F = a(T)|\Phi|^2 + \frac{b}{2}|\Phi|^4 + T_0 \int \frac{d^3\vec{q}}{(2\pi)^3} \ln(q^2 + H + 2e^2|\Phi|^2),$$

so that after the shift in transition temperature

$$\frac{dF}{d|\Phi|} = 2a(T)|\Phi| + 2b|\Phi|^3 - \frac{2e^4 T_0}{\pi} \frac{|\Phi|^3}{\sqrt{H} + \sqrt{H + 2e^2|\Phi|^2}}.$$

By expanding in Φ and then integrating, the free energy becomes

$$F = a(T)\Phi^2 + \frac{1}{2}\left(b - \frac{e^4 T_0}{2\pi H^{1/2}}\right)\Phi^4 + \frac{e^6}{12\pi H^{3/2}}\Phi^6 + O(\Phi^8).$$

For the transition to be discontinuous the coefficient of Φ^4 in F has to be negative. The critical field where the transition becomes continuous is therefore

$$H_c = \left(\frac{e^4 T_0}{2\pi b}\right)^2.$$

Problem 4.4 Show that the size of the first-order transition t_{fo} in Eq. (4.16) is proportional to the ratio of the latent heat and the mean-field discontinuity of the specific heat at the transition for $e = 0$.

Solution The latent heat per unit volume at the first order transition is $\Delta Q = T_c \partial F/\partial T = a_0 T_c \Phi^2$. Since $\Phi^2 = (\sqrt{2}e^3 T_0/3\pi b)^2$ at the first-order transition, and the specific heat discontinuity at the $e = 0$ transition is $\Delta C = |a(0)|^2/bT_0$, it follows that

$$\frac{\Delta Q}{T_c \Delta C} = \frac{2e^6 T_0^2}{(3\pi)^2 b|a(0)|} = 2t_{fo}.$$

4.3 Type-II superconductors near four dimensions

In type-II superconductors fluctuations of the order parameter cannot be neglected, and the mean-field theory from the previous section is inconsistent. We may therefore try the renormalization group approach in which one would integrate over both the order parameter and the vector field modes with the large wavevectors $\Lambda/s < |\vec{k}| < \Lambda$. Such an integration will generate flows of the three couplings: $a(T)$, b, and the charge e. The result of the

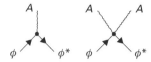

Figure 4.3 Gauge field–order parameter interaction vertices in Eq. (4.22).

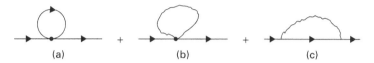

Figure 4.4 Diagrams contributing to the renormalization of the quadratic part in Φ of S_{GL}, to the lowest order in quartic interaction and the charge.

momentum-shell integration is the action for the slow modes:

$$
\begin{aligned}
S_{GL} = {} & \int_0^{\Lambda/s} \frac{d\vec{k}}{(2\pi)^d}[(Z_\eta k^2 + Z_a a(T))\Phi^*(\vec{k})\Phi(\vec{k}) \\
& + Z_{v1}e \int_0^{\Lambda/s} \frac{d\vec{q}}{(2\pi)^d}(2\vec{k}+\vec{q})\cdot\vec{A}(\vec{q})\Phi^*(\vec{k}+\vec{q})\Phi(\vec{k}) \\
& + Z_{v2}e^2 \int_0^{\Lambda/s} \frac{d\vec{q}\,d\vec{p}}{(2\pi)^{2d}}\vec{A}(\vec{q})\cdot\vec{A}(\vec{p})\Phi^*(\vec{k}+\vec{q}+\vec{p})\Phi(\vec{k})] \\
& + \frac{b'}{2}\int_0^{\Lambda/s}\frac{d\vec{k}_1...d\vec{k}_4}{(2\pi)^{3d}}\Phi^*(\vec{k}_1)\Phi^*(\vec{k}_2)\Phi(\vec{k}_3)\Phi(\vec{k}_4)\delta(\vec{k}_1+\vec{k}_2-\vec{k}_3-\vec{k}_4) \\
& + \frac{1}{2}\int_0^{\Lambda/s}\frac{d\vec{k}}{(2\pi)^2}\left[Z_A k^2(\delta_{ij}-\hat{k}_i\hat{k}_j) + \frac{k_i k_j}{g}\right]A_i(\vec{k})A_j(-\vec{k}). \quad (4.22)
\end{aligned}
$$

Besides the quartic interaction, when $e \neq 0$ there are two more interaction vertices corresponding to the second and the third terms in the above equation, diagrammatically depicted in Fig. 4.3. We consider the general number of complex order parameter components N throughout the calculation, and assume $d = 4 - \epsilon$ dimensions. We have also introduced the "gauge-fixing parameter" g. Taking $g = 0$ completely suppresses the longitudinal component of the vector potential, and therefore defines the transverse gauge used in the previous section. The reader may wish to consult Appendix C about the gauge-fixing procedure and this particular choice of gauge.

The renormalization factors Z_η and Z_a follow from the diagrams in Fig. 4.4. Diagram 4.4 (a) has already been calculated in Problem 3.1, and yields $Z_a = 1 - (N + 1)\hat{b}\ln(s)$. Diagram 4.4 (b) gives only a finite shift of the transition

Figure 4.5 Lowest order contributions to Z_{v1}.

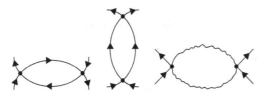

Figure 4.6 Lowest order contributions to the renormalized quartic interaction b.

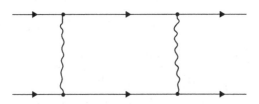

Figure 4.7 An example of vanishing diagram at zero external wavevectors.

temperature. Diagram 4.4 (c), on the other hand, yields

$$-e^2 \int_{\Lambda/s}^{\Lambda} \frac{d\vec{q}}{(2\pi)^d} \frac{(2k+q)_i(\delta_{ij} - \hat{q}_i\hat{q}_j)(2k+q)_j}{q^2((\vec{q}+\vec{k})^2 + a(T))}$$
$$= -3k^2\hat{e}^2 \ln(s) + O(k^4, k^2 a(T)), \tag{4.23}$$

and therefore contributes to Z_η: $Z_\eta = 1 - 3\hat{e}^2 \ln(s)$. Here $\hat{e}^2 = e^2 \Lambda^{d-4} S_d / (2\pi)^d$ is the dimensionless charge.

The reader is invited to compute the renormalization factors Z_{v1} and Z_{v2}, and show that to the order e^2, $Z_{v1} = Z_{v2} = Z_\eta$. Diagrams contributing to Z_{v1}, for example, are presented in Fig. 4.5. This is an example of a *Ward–Takahashi identity* which holds to all orders in perturbation theory, and is a consequence of the gauge invariance of the Ginzburg–Landau action. It implies that the Ginzburg–Landau action retains its gauge-invariant form in Eq. (4.1) as the ultraviolet cutoff is changed.

The coupling b' follows from the diagrams in Fig. 4.6. All other diagrams that could potentially contribute vanish in the transverse gauge when the external momenta are set to zero. An example of such a diagram is provided in Fig. 4.7. The only finite new diagram is then the last one in Fig. 4.6. It is

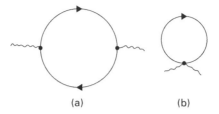

Figure 4.8 Lowest-order polarization diagrams determining the factor Z_A.

easily computed to be

$$-e^4 \int_{\Lambda/s}^{\Lambda} \frac{d\vec{q}}{(2\pi)^d} \frac{(\delta_{ij} - \hat{q}_i \hat{q}_j)(\delta_{ij} - \hat{q}_i \hat{q}_j)}{q^4} = -3e^2 \hat{e}^2 \ln(s), \qquad (4.24)$$

in four dimensions. All three diagrams together then give

$$\hat{b}' = \hat{b} - (N + 4)\hat{b}^2 \ln(s) - 6\hat{e}^4 \ln(s). \qquad (4.25)$$

Finally, the "polarization" diagram in Fig. 4.8 (a) gives

$$-Ne^2 \int_{\Lambda/s}^{\Lambda} \frac{d\vec{q}}{(2\pi)^d} \frac{(\vec{k} + 2\vec{q})_i (\vec{k} + 2\vec{q})_j}{(q^2 + a(T))((\vec{q} + \vec{k})^2 + a(T))}$$

$$= \frac{N\hat{e}^2}{3} \ln(s)(k^2 \delta_{ij} - k_i k_j), \qquad (4.26)$$

which yields $Z_A = 1 + N\hat{e}^2 \ln(s)/3$. Notice that the polarization is purely transverse, so that the gauge-fixing parameter g is not renormalized. In the result in Eq. (4.26) we have discarded the spurious wavevector-independent term as an artifact of the ultraviolet cutoff, which evidently breaks the gauge invariance at large wavevectors. The diagram in Fig. 4.8 (b) is dropped for the same reason.[2]

We can now rescale the momenta and the fields and read off the renormalized parameters in the action for the slow modes. As before, $a(T, s) = Z_a a(T)s^2/Z_\eta$ and $b(s) = b's^\epsilon/Z_\eta^2$. After redefining the vector potential as $e\vec{A} \to \vec{A}$, we can define the renormalized charge as $e^2(s) = e^2 s^\epsilon / Z_A$. Differentiating with respect to $\ln(s)$, one obtains the requisite β-functions to the lowest order in couplings b and e and parameter ϵ,

$$\frac{da(T)}{d\ln(s)} = a(T)(2 - (N + 1)b + 3e^2), \qquad (4.27)$$

---

[2] The field-theoretic dimensional regularization discussed in Sections 3.6 and 3.7 does not require the cutoff, and is designed to automatically preserve gauge invariance of the theory.

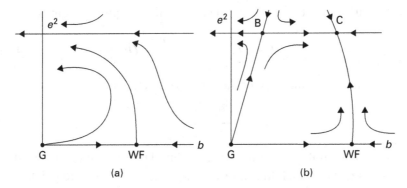

Figure 4.9 The flow of the quartic coupling b and the charge e in the critical plane $a(T) = 0$, below N_c (a) and above N_c (b).

and

$$\frac{db}{d\ln(s)} = \epsilon b + 6e^2 b - (N+4)b^2 - 6e^4, \tag{4.28}$$

$$\frac{de^2}{d\ln(s)} = \epsilon e^2 - \frac{N}{3}e^4, \tag{4.29}$$

where the hats on the dimensionless couplings b and e^2 have been dropped.

The structure of the flow diagram in the critical plane $a(T) = 0$ depends crucially on the number of components N. There are always the usual Gaussian and the Wilson–Fisher fixed points at $e^* = 0$, which are both unstable in the charge direction. The other fixed points lie at $(e^*)^2 = 3\epsilon/N$ and

$$b_{\pm}^* = \frac{\epsilon}{2(N+4)}\left(1 + \frac{18}{N} \pm \sqrt{\left(1 + \frac{18}{N}\right)^2 - \frac{216(N+4)}{N^2}}\right). \tag{4.30}$$

$b_{\pm}^*$ are real for $N > N_c = 182.952$, and complex otherwise. For $N > N_c$ therefore, besides the Gaussian and the Wilson–Fisher fixed points there are two more *charged* fixed points, one of which is stable, and the other unstable in the b direction. The former is the critical point, and the latter *bicritical*, since it has two unstable directions. The line that connects the Gaussian and the bicritical point separates the domain of attraction of the critical point from the region where the flow is towards $b < 0$. When $b < 0$ one needs to add a positive $|\Phi|^6$ term in the Ginzburg–Landau action, and this generically leads to a first order transition (see Problem 4.2). For $N < N_c$, on the other hand, there are no charged fixed points at real values of the interaction b and the flow is always towards the region $b < 0$, and the transition is first order. The two qualitatively different situations are depicted in Fig. 4.9.

At the critical point for $N > N_c$ the correlation length exponent is

$$v^{-1} = 2 - (N + 1)b_+^* + \frac{9\epsilon}{N} + O(\epsilon^2), \tag{4.31}$$

and the anomalous dimension of the order parameter is

$$\eta = -\frac{9\epsilon}{N} + O(\epsilon^2). \tag{4.32}$$

Other exponents follow from the usual scaling laws, which are all satisfied, Josephson's law included. In contrast to the Wilson–Fisher fixed point, at the charged critical point the anomalous dimension is finite (and negative) already to the lowest order in ϵ.

It therefore appears that for $N = 1$ the superconducting transition remains discontinuous even after the inclusion of the fluctuations of the order parameter. The large number for N_c is deceiving, however. To the next order in ϵ-expansion one finds an equally large and negative correction to N_c (Problem 4.6), which suggests that in three dimensions N_c may in fact be drastically reduced. Today it is believed that for $N = 1$ and in $d = 3$ the flow diagram is in fact qualitatively the same as that in Fig. 4.9 (b), the lowest-order ϵ-expansion result notwithstanding. We will return to this issue in Chapter 7 when we introduce the concept of duality.

Problem 4.5 Find the values of the Ginzburg–Landau parameter κ which remain invariant under the change of cutoff. What is their physical significance?

Solution Since $\kappa^2 = b/2e^2$, Eqs. (4.28) and (4.29) imply

$$\frac{d\kappa^2}{d \ln(s)} = e^2 \left(-3 + 2 \left(3 + \frac{N}{6} \right) \kappa^2 - 2(N + 4)\kappa^4 \right).$$

The Ginzburg–Landau parameter remains constant during renormalization for

$$\kappa_\pm^2 = \frac{1}{2(N + 4)} \left[\left(3 + \frac{N}{6} \right) \pm \sqrt{\left(3 + \frac{N}{6} \right)^2 - 6(N + 4)} \right].$$

Again, the solutions are real only for $N > 182.952$. For large N,

$$\kappa_-^2 = \frac{9}{N} + O \left(\frac{1}{N^2} \right).$$

and

$$\kappa_+^2 = \frac{1}{6} + O\left(\frac{1}{N}\right).$$

For $\kappa < \kappa_-$ the Ginzburg–Landau parameter decreases under renormalization and the transition is discontinuous, while for $\kappa > \kappa_-$ the transition is continuous. κ_+ determines the universal ratio of the penetration depth and the correlation length near the critical point.

Problem 4.6* Find the first-order correction in ϵ to the value of N_c using the next-order β-functions:

$$\beta_e = \epsilon e^2 - \frac{N}{3}e^4 - 2Ne^6,$$

$$\beta_b = \epsilon b + 6e^2 b - (N+4)b^2 - 6e^4 + \frac{3(3N+7)}{2}b^3$$

$$- 2(2N+5)b^2 e^2 - \frac{71N+87}{6}be^4 + \frac{2(7N+45)}{3}e^6.$$

Solution From $\beta_e = 0$ it follows that $(e^*)^2 = (3\epsilon/N) - (54\epsilon^2/N^2) + O(\epsilon^3)$. Inserting that into $\beta_b = 0$ and denoting $b^* = x\epsilon$, one finds the equation for x:

$$x\left(1 + \frac{18}{N} - \left(\frac{18}{N}\right)^2 \epsilon - \frac{3(71N+87)}{2N^2}\epsilon\right) - x^2\left(N + 4 + \frac{6(2N+5)}{N}\epsilon\right)$$

$$+ \frac{3(3N+7)}{2}\epsilon x^3 - \frac{54}{N^2} + \frac{18(7N+153)}{N^3}\epsilon = 0.$$

For $\epsilon = 0$ the quadratic equation for x has two real roots for $N > N_{c0}$, which merge at $N = N_{c0}$, and become complex for $N < N_{c0}$. At $N = N_{c0}$ and $\epsilon = 0$ the above equation may therefore be written as

$$(N_{c0} + 4)(x - x_{c0})^2 = 0.$$

Matching the coefficients yields

$$x_{c0} = \frac{N_{c0} + 18}{2N_{c0}(N_{c0} + 4)}$$

and

$$N_{c0} = 90(1 + \sqrt{1 + (1/15)}) = 182.952.$$

Similarly, for $\epsilon \neq 0$ the cubic equation for x at $N = N_c$ can be written as

$$\frac{3\epsilon}{2}(3N_c + 17)(x - x_0)(x - x_c)^2 = 0,$$

where

$$x_0 = \frac{2(N_c + 4)}{3(3N_c + 7)\epsilon} + O(1)$$

is the unphysical real root that exists at all N, and $x_c = x_{c0} + O(\epsilon)$. Matching the coefficients again gives three equations for x_0, x_c and N_c. Numerically solving for N_c at small ϵ gives

$$N_c = N_{c0}(1 - 1.75\epsilon + O(\epsilon^2)).$$

4.4 Anomalous dimension for the gauge field

In addition to the standard critical exponents introduced in Chapter 1, with the fluctuating vector potential one may define one more exponent. For $T < T_c$ the gauge-field correlation function in the transverse gauge can be written in the scaling form

$$\langle A_i(\vec{q})A_j(-\vec{q})\rangle = (\delta_{ij} - \hat{q}_i\hat{q}_j)q^{-(2-\eta_A)}F_-(q\lambda), \tag{4.33}$$

where η_A is the anomalous dimension of the gauge field, and λ the magnetic field penetration depth. The scaling function $F_-(z)$ approaches a constant when z tends to infinity, that is right at the critical point, and $F_-(z) \sim z^{2-\eta_A}$ for z small. An analogous scaling form may be written for $T > T_c$, with a different function $F_+(q\xi)$, and with the correlation length ξ replacing the penetration depth λ, the latter being infinite.

Interestingly, the exponent η_A can be obtained exactly. In analogy to the anomalous dimension for the order parameter η, η_A is defined as

$$\eta_A = \frac{d \ln Z_A}{d \ln(s)}\Big|_{e=e^*,b=b^*}. \tag{4.34}$$

Since $e^2(s) = s^\epsilon e^2/Z_A$, differentiating with respect to $\ln(s)$ we can write

$$\beta_e = e^2(\epsilon - \eta_A). \tag{4.35}$$

When $e^* = 0$, like at the Wilson–Fisher fixed point, $Z_A = 0$ and consequently $\eta_A = 0$. At the charged fixed points, however, $e^* \neq 0$ and $\eta_A = \epsilon$, exactly, with no further corrections. In particular, in $d = 3$, $\eta_A = 1$.

The physical meaning of this exact result may be understood as follows. Recall that in order to determine the renormalization of the charge we had to redefine the vector potential in Eq. (4.2) as $e\vec{A} \to \vec{A}$. The new vector potential then has the dimension of inverse of length, i.e. $A \sim \xi^{-1}$. The gauge invariance of Eq. (4.2) implies that this simple result of dimensional analysis is in fact exact. Combined with the hyperscaling relation $L_{GL} \sim \xi^{-d}$, this implies that the Higgs mass of the gauge-field in the ordered phase scales like $\sim \xi^{2-d}$. Finally, from the scaling form in Eq. (4.33) that same mass behaves like $\lambda^{\eta_A - 2}$. This implies that

$$\lambda^{\eta_A - 2} \propto \xi^{2-d}. \tag{4.36}$$

Since $\eta_A = 4 - d$ exactly, in any dimension

$$\lambda \propto \xi, \tag{4.37}$$

in the critical region. The value of the anomalous dimension η_A ensures that the two characteristic lengths diverge with precisely the same exponent ν, just like the mean-field theory would suggest. The Ginzburg–Landau parameter becomes therefore a finite universal number that characterizes the charged critical point (Problem 4.5).

Right at $T = T_c$, $\langle A_i(\vec{q})A_j(-\vec{q})\rangle \sim q^{-1}$ in $d = 3$, which implies $\sim r^{-2}$ in real space. This qualitative change in correlations of the gauge-field at the charged critical point from $\sim r^{-1}$ to $\sim r^{-2}$ has been confirmed in numerical calculations.

4.5 Width of the critical region

The effects of the charged critical point may be expected to become observable in type-II materials, such as high-temperature superconductors. The width of the critical region may be estimated from requiring that the dimensionless charge approaches its fixed point value, $e^2 \xi \sim 1$. Recalling that $2e^2 = \kappa^{-2}b$, $\xi \sim \sqrt{t}$, and that the width of the Wilson–Fisher critical region $t_{cr} \sim b^2$, we find

$$t_{charged} \sim t_{cr}\kappa^{-4}. \tag{4.38}$$

In low-temperature superconductors like elemental metals the width of the critical region associated with the Wilson–Fisher fixed point is $t_{cr} \sim 10^{-16}$, and $t_{charged}$ is likewise immeasurably small. In high-temperature superconductors $t_{cr} \sim 10^{-1}$, and the effects of order-parameter fluctuations are readily observable. On the other hand, a typical value of the Ginzburg–Landau parameter is $\kappa^2 \sim 10^3$, so $t_{charged} \sim 10^{-7}$, which is again too small to be easily observable.

Problem 4.7 What value of the critical exponent for the divergence of the penetration depth do you expect to be observable in high-temperature superconductors?

Solution Since in high-temperature superconductors $t_{cr} \propto 10^{-1}$ and $\kappa^2 \sim 10^3$, the scaling for the reduced temperatures $10^{-1} < t < 10^{-3}$ is governed by the flow near the unstable Wilson–Fisher fixed point in three dimensions, with $\eta_A = 0$. Eq. (4.36) implies then that

$$\lambda \propto t^{-\nu/2},$$

with $\nu \approx 0.67$ corresponding to $N = 2$ and $d = 3$ in Table 3.1.

Problem 4.8 Determine the gauge-field correlation function at $T > T_c$ in $d = 3$ using the perturbation theory to the first order in e^2.

Solution To the lowest order in charge the polarization is given by the diagrams in Fig. 4.8. This yields

$$\Pi(k) = N e^2 \int \frac{d\vec{q}}{(2\pi)^3} \left[\frac{2\delta_{ij}}{q^2 + a(T)} - \frac{(2q + k)_i (2q + k)_j}{((\vec{q} + \vec{k})^2 + a(T))(q^2 + a(T))} \right].$$

After rewriting the denominator in the second integral using Feynman's identity in Eq. (3.77),

$$\frac{1}{AB} = \int_0^1 \frac{dx}{(xA + (1 - x)B)^2},$$

and then making the change of variables from $\vec{q}$ to $\vec{p} = \vec{q} + x\vec{k}$, the integration over $\vec{p}$ and the parameter x gives

$$\Pi(k) = \frac{e^2 N}{8\pi} k \left(\left(1 + \left(\frac{2}{\xi k} \right)^2 \right) \arctan \left(\frac{\xi k}{2} \right) - \frac{2}{\xi k} \right) (\delta_{ij} - \hat{k}_i \hat{k}_j),$$

where $a(T) = \xi^{-2}$. The gauge-field correlation function is therefore

$$\langle A_i(\vec{k}) A_j(-\vec{k}) \rangle = (k^2 + \Pi(k))^{-1} (\delta_{ij} - \hat{k}_i \hat{k}_j).$$

In the critical region when $\xi k \gg 1$, $\Pi(k) = e^2 N k / 16$, whereas for $\xi k \ll 1$ one finds $\Pi(k) = e^2 N \xi k^2 / (24\pi)$.

Problem 4.9* Determine the effect of order parameter fluctuations on the size of the first-order transition in good type-I superconductors ($\kappa \ll 1$) with small charge, for $N \gg 1$.

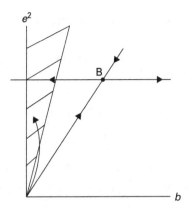

Figure 4.10 The renormalization trajectory of a good type-I material. The mean-field result for the transition temperature in Eq. (4.15) applies in the shaded region.

Solution Assume that the transition with the original values of the couplings e and b occurs at some reduced temperature t. The change of the cutoff $\Lambda \to \Lambda/s$ is equivalent to changing the coupling constants to $e(s)$ and $b(s)$ (with the hats being omitted), according to the differential equations (4.28)–(4.29). At some $s > 1$, the first-order transition with the new couplings would occur at some t_0. If t_0 is large enough, we may approximate it with the mean-field result

$$t_0 = \frac{e^6(s)}{(3\pi)^2 a_0 b(s)},$$

as in Eq. (4.15), where we set $T_0 = 1$ for convenience. This equation together with the boundary conditions $t(1) = t$ and $t(s) = t_0$ then determines the actual size of the transition t.

In the zeroth-order approximation in the coupling constants, the relations $t_0/t = s^2$, $e^2(s) = e^2 s$ and $b(s) = bs$ in $d = 3$ correctly give $t = t_{\mathrm{MF}} = e^6/((3\pi)^2 a_0 b)$. For $\kappa \ll 1$, $b(s)$ starts decreasing while the other two couplings still scale as $e^2(s) \approx e^2 s$ and $t(s) = ts^2$ (Fig. 4.10). We may then write

$$a_0 t s^2 = \frac{e^4 s^2}{2(3\pi)^2 \kappa^2(s)},$$

or, eliminating the charge,

$$\frac{t}{t_{\mathrm{MF}}} = \frac{\kappa^2}{\kappa^2(s)},$$

where $s = \sqrt{t_0/t}$. Using the result that for small κ

$$\frac{d\kappa^2(s)}{d\ln(s)} = -3e^2(s),$$

one finds

$$\frac{t}{t_{\mathrm{MF}}} = 1 + \frac{3e^2}{\kappa^2}\left(\frac{t_0}{t_{\mathrm{MF}}}\right)^{1/2}.$$

Problem 4.10* Determine the size of the first order transition close to the separatrix between the type-I and type-II regions in $4 - \epsilon$ dimensions and for $N > N_{\mathrm{c}}$.

Solution Near and left of the separatrix we can assume that $e(s) \approx e^*$ while $b^* - b(s) = (b^* - b)s^\phi$, where the *crossover exponent* ϕ is

$$\phi = \epsilon\sqrt{\left(1 + \frac{18}{N}\right)^2 - \frac{216(N+4)}{N^2}}.$$

Since $t_0/t = s^{1/\nu_-}$, with $\nu_-^{-1} = 2 - (N+1)b_-^* + 9\epsilon/N$ at the bicritical point, one finds

$$t \sim (b_{\mathrm{sep}} - b)^{\frac{1}{\phi\nu}},$$

where b_{sep} is the charge-dependent value of b at the separatrix between the type-I and the type-II regions.

Problem 4.11* Find the size of the first-order transition for $N < N_{\mathrm{c}}$ in $4 - \epsilon$ dimensions for small charge.

Solution Similar to the solution of the previous problem,

$$t \sim e^{2/(\phi\nu_{\mathrm{WF}})},$$

where ν_{WF} is the correlation length exponent and $\phi = \epsilon$ is the second positive eigenvalue of the stability matrix at the unstable Wilson–Fisher fixed point.

5

Near lower critical dimension

The destruction of long-range order in sufficiently low dimensions due to fluctuations of Goldstone's modes is described. The Mermin–Wagner–Hohenberg theorem is proven for the XY model. The non-linear σ-model for interacting Goldstone modes is formulated and studied near two dimensions and at low temperatures. Renormalization of the non-linear σ-model is discussed.

5.1 Goldstone modes

We now return to the neutral superfluid and magnetic transitions, and take the limit $e = 0$ in the Ginzburg–Landau action in Eq. (4.2). In the ordered phase, expansion around the saddle-point value of the field in terms of the amplitude and phase fluctuations led to Eq. (4.4). We already noted that whereas the amplitude fluctuations were massive the phase fluctuations were massless; their energy vanished as the wavevector approached zero. This has profound consequences for the ordering below the transition temperature. Consider the thermal average

$$\langle \phi(\vec{x})\phi(\vec{y}) \rangle = \frac{T}{2|\Phi_0|^2} \int \frac{d\vec{k}}{(2\pi)^d} \frac{e^{i\vec{k}\cdot(\vec{x}-\vec{y})}}{k^2}, \tag{5.1}$$

in the Gaussian approximation in Eq. (4.4), where as always, we are assuming a finite ultraviolet cutoff on the momentum integral. For $\vec{x} = \vec{y}$ the above integral is divergent for dimensions $d \leq 2$, and finite otherwise. This divergence obviously may invalidate our assumption that the ordered state in $d \leq 2$ existed in the first place. Let us assume a low temperature, so that the fluctuations of the amplitude of the order parameter are negligible. We may

97

then write

$$\langle \Phi^*(\vec{x})\Phi(0)\rangle \approx |\Phi_0|^2 \langle e^{i(\phi(\vec{x})-\phi(0))}\rangle. \tag{5.2}$$

The average on the right hand side, on the other hand, in the Gaussian approximation may be written as

$$\langle e^{i(\phi(\vec{x})-\phi(0))}\rangle = e^{-\frac{1}{2}\langle(\phi(\vec{x})-\phi(0))^2\rangle} = \exp\left(-\frac{T}{2|\Phi_0|^2}\int \frac{d\vec{k}}{(2\pi)^d}\frac{1-e^{i\vec{k}\cdot\vec{x}}}{k^2}\right). \tag{5.3}$$

The last equation may be checked directly by expanding the average on the left hand side and using Wick's theorem for Gaussian averages. The integral in the exponent is finite when $d > 2$. In $d = 2$ however, it is

$$\int_0^{\Lambda x} \frac{dq}{2\pi q}\int_0^{2\pi}\frac{d\theta}{2\pi}(1-e^{iq\cos\theta}) = \int_0^{\Lambda x}\frac{dq}{2\pi q}(1-J_0(q)), \tag{5.4}$$

where the Bessel function $J_0(q) \approx (2/\pi q)^{1/2}\cos(q-(\pi/4))$ for $q \gg 1$. In $d = 2$ therefore

$$\langle \Phi^*(\vec{x})\Phi(0)\rangle \approx |\Phi_0|^2 e^{-\frac{T\ln(\Lambda x)}{4\pi|\Phi_0|^2}} \sim x^{-\frac{T}{4\pi|\Phi_0|^2}}, \tag{5.5}$$

and the correlation function approaches zero at large separations, albeit slowly, even at low temperatures. Fluctuations of the phase have destroyed long-range order in $d = 2$ at any finite temperature! The effect is even stronger in $d < 2$. The value of the dimension below which there is no long-range order in a given model will be called the *lower critical dimension*, d_{low}. In the above example $d_{low} = 2$.

Although the above simple argument suggests that the superfluid long-range order in $d = 2$ is not possible, the power-law decay of the correlation function is still qualitatively different from the exponential behavior expected at high temperatures. We will therefore call the phase with such power-law correlations *algebraically ordered*. In the case of a superfluid in $d = 2$ there exists a special transition between the low temperature state with the algebraic order and the exponentially disordered high-temperature state, which will be the subject of the next chapter.

The phase-like massless modes always exist when the ordered state breaks a continuous global symmetry of the action. The rigorous version of this statement is called the *Nambu–Goldstone theorem*, and the massless modes, the *Goldstone modes*. The number of Goldstone modes equals the number of broken generators of the symmetry group. For example, for the Heisenberg model the symmetry of the action is the group of rotations in three dimensional order-parameter space $O(3)$. In the ordered state this becomes broken down to $O(2)$, the rotations around the axis of finite magnetization. Since the rotations around the two remaining orthogonal axis are not the symmetry operations in the ordered state, there are then *two* Goldstone modes. For the Ising model, on the other hand, the global symmetry is the discrete transformation Z_2 that exchanges "up" with "down", and there are no Goldstone modes. As a result $d_{\text{low}} = 1$ for the Ising model (Problem 1.1), while $d_{\text{low}} = 2$ for the $O(N)$ model.

The statement that Goldstone modes tend to destroy long-range order at finite temperatures in $d = 2$ may be rigorously proven for a general class of models with global $O(N)$ symmetry and with sufficiently short-ranged interactions, and is known as the *Mermin–Wagner–Hohenberg theorem*, discussed in the next section. This does not necessarily mean that there cannot be a phase transition in $d = 2$, and, as mentioned above, indeed there is one when the symmetry is $O(2)$.

The Nambu–Goldstone theorem breaks down when there exist sufficiently long-ranged interactions in the system, like the one mediated by the gauge-field in the previous chapter. In that case, what would be a massless Goldstone mode becomes pushed to a finite energy as $k \to 0$ by the Higgs mechanism.

5.2 Mermin–Wagner–Hohenberg theorem

Here we provide the proof of the Mermin–Wagner–Hohenberg theorem, that states that there cannot be true long-range order at any finite temperature in two-dimensional systems with continuous symmetry. For brevity we will consider only the simplest such system, the XY model in Eq. (1.4) with $O(2)$ symmetry, but the proof directly generalizes to $O(N)$ symmetry.[1]

Energy of an arbitrary configuration of the XY model in the uniform magnetic field from Eq. (1.4) is

$$E = -J \sum_{\langle i,j \rangle} \cos(\phi_i - \phi_j) - H \sum_{i=1}^{N} \cos \phi_i. \tag{5.6}$$

where ϕ_i is the angle of the unit vector $\vec{s}_i$ at a lattice site $\vec{r}_i$ measured from the direction of the field $\vec{H}$. We will assume a two-dimensional quadratic lattice

---

[1] I am indebted to Professor H. Nishimori for this version of the proof.

with N sites and with the lattice constant set to unity. The proof relies on the *Bogoliubov–Schwartz inequality*:

$$\langle A A^* \rangle \geq \frac{|\langle A B^* \rangle|^2}{\langle B B^* \rangle}. \tag{5.7}$$

The inequality is valid for arbitrary quantities A and B, and the symbol $\langle \ldots \rangle$ denotes the usual statistical average.

The Mermin–Wagner–Hohenberg theorem then follows from a judicious choice of A and B:

$$A = \frac{1}{N} \sum_j e^{-i\vec{q} \cdot \vec{r}_j} \sin \phi_j, \tag{5.8}$$

$$B = \frac{1}{N} \sum_k e^{-i\vec{q} \cdot \vec{r}_k} \frac{\partial E}{\partial \phi_k}. \tag{5.9}$$

First, note that from the definition of A,

$$\sum_{\vec{q}} \langle A A^* \rangle = \frac{1}{N} \sum_j \langle \sin^2 \phi_j \rangle \leq 1. \tag{5.10}$$

Similarly,

$$\langle A B^* \rangle = \frac{1}{N^2} \sum_{j,k} e^{-i\vec{q} \cdot (\vec{r}_j - \vec{r}_k)} \left\langle \frac{\partial E}{\partial \phi_k} \sin \phi_j \right\rangle, \tag{5.11}$$

where the average on the right hand side can be written as

$$\left\langle \sin \phi_j \frac{\partial E}{\partial \phi_k} \right\rangle = \frac{1}{Z} \int_0^{2\pi} \left(\prod_i d\phi_i \right) e^{-\frac{E}{T}} \frac{\partial E}{\partial \phi_k} \sin \phi_j = \delta_{jk} T \langle \cos \phi_j \rangle. \tag{5.12}$$

The result follows from integrating by parts and using the periodicity of the XY model. It follows then that

$$\langle A B^* \rangle = \frac{T m}{N}, \tag{5.13}$$

where $m = \langle \cos \phi_k \rangle$ is the magnetization.

Finally,

$$\langle B B^* \rangle = \frac{T}{N^2} \sum_{j,k} e^{-i\vec{q} \cdot (\vec{r}_k - \vec{r}_j)} \left\langle \frac{\partial^2 E}{\partial \phi_k \partial \phi_j} \right\rangle, \tag{5.14}$$

where we have again used an integration by parts. Taking the derivatives after some algebra yields

$$\langle BB^* \rangle = \frac{TJ}{N^2} \sum_k \sum_j{}' \left(1 - e^{-i\vec{q}\cdot(\vec{r}_k - \vec{r}_j)} \right) \langle \cos(\phi_k - \phi_j) \rangle$$
$$+ \frac{TH}{N^2} \sum_l \langle \cos \phi_l \rangle, \tag{5.15}$$

where the prime on the sum means that the summation over j is restricted to the nearest neighbors of the site k. Since the average $\langle \cos(\phi_k - \phi_j) \rangle$ is actually a constant for all nearest neighbors k and j, we can also rewrite the last expression as

$$\frac{\langle BB^* \rangle}{T} = \frac{J}{N^2} \sum_k \left(4 - \sum_j{}' e^{-i\vec{q}\cdot(\vec{r}_k - \vec{r}_j)} \right) \langle \cos(\phi_k - \phi_j) \rangle + \frac{H}{N^2} \sum_l \langle \cos \phi_l \rangle, \tag{5.16}$$

keeping in mind that the site j outside the sum is a nearest neighbor of k. So,

$$\frac{\langle BB^* \rangle}{T} = \frac{J}{N}(4 - 2\cos q_x - 2\cos q_y)\langle \cos(\phi_k - \phi_j) \rangle + \frac{H}{N} \langle \cos \phi_l \rangle, \tag{5.17}$$

and therefore

$$\langle BB^* \rangle \le \frac{T\left(H + J\left(q_x^2 + q_y^2 \right) \right)}{N}. \tag{5.18}$$

Combining the results (5.10), (5.13), and (5.18) with the Bogoliubov–Schwartz inequality then gives

$$1 \ge \sum_{\vec{q}} \langle AA^* \rangle \ge \sum_{\vec{q}} \frac{(Tm/N)^2}{\langle BB^* \rangle} \ge \frac{Tm^2}{N} \sum_{\vec{q}} \frac{1}{H + Jq^2}. \tag{5.19}$$

Taking the thermodynamic limit $N \to \infty$ implies that

$$1 \ge Tm^2 \int \frac{d^2\vec{q}}{(2\pi)^2} \frac{1}{H + Jq^2}. \tag{5.20}$$

Since the integral on the right hand side diverges as the magnetic field diminishes, we proved that in the thermodynamic limit,

$$\lim_{H \to 0} m = 0 \tag{5.21}$$

in two dimensions. Notice that it is the same logarithmic infrared divergence as in Eq. (5.5) that is responsible for zero magnetization. Obviously the divergence is only stronger in one dimension, and the theorem likewise applies there as well. It does not exclude an algebraic order, nor does it readily apply to systems with long-ranged interactions. Some examples that follow in the latter category are provided in Problems 5.5 and 5.6.

5.3 Non-linear σ-model

Let us then consider the transition in the Ginzburg–Landau–Wilson theory with $O(N)$ symmetry starting from the ordered phase. For convenience, let us assume a magnetic order parameter, and consider a lattice with a small but finite lattice spacing a. We will also find it useful to add a small magnetic field $\vec{B}$ which explicitly breaks $O(N)$ invariance and selects the axis of magnetization. The field $\Phi(\vec{x}) = (\sigma(\vec{x}), \vec{\Pi}(\vec{x}))$, where $\sigma(\vec{x})$ is along the magnetic field, and $\vec{\Pi}(\vec{x})$ has $N - 1$ real components orthogonal to $\vec{B}$: $\vec{B} \cdot \vec{\Pi}(\vec{x}) = 0$. If we assume sufficiently low temperatures and neglect the amplitude fluctuations, we may write the constraint on the field components as

$$\sigma^2(\vec{x}) + \vec{\Pi}^2(\vec{x}) = \Phi_0^2, \tag{5.22}$$

and set $\Phi_0^2 = 1$ for convenience. The partition function may then be written as

$$Z = \int D\vec{\Pi}(\vec{x}) D\sigma(\vec{x}) \prod_{\vec{x}} \delta(\sigma^2(\vec{x}) + \vec{\Pi}^2(\vec{x}) - 1)$$

$$\times e^{-\frac{1}{T} \int d\vec{x} [\frac{1}{2}((\nabla\sigma(\vec{x}))^2 + (\nabla\vec{\Pi}(\vec{x}))^2) - B\sigma(\vec{x})]}, \tag{5.23}$$

which defines the *non-linear σ-model*. In Eq. (5.23) we neglected terms of higher order in fields and their derivatives. Although the action is then quadratic in the fields σ and $\vec{\Pi}$, the model is not simply Gaussian because of the constraint in the measure of the integral. The partition function in Eq. (5.23) is presumed to describe the same phase transition as in the original Ginzburg–Landau–Wilson theory, only this time from the low-temperature side.

If the constraint is solved, and $\sigma(\vec{x}) = \sqrt{1 - \vec{\Pi}^2(\vec{x})}$ substituted into the partition function, Z becomes

$$Z = \int \prod_{\vec{x}} \frac{d\vec{\Pi}(x)}{2\sqrt{1 - \vec{\Pi}^2(\vec{x})}} e^{-\frac{1}{T} \int d\vec{x} [\frac{1}{2}((\nabla\sqrt{1 - \vec{\Pi}^2(\vec{x})})^2 + (\nabla\vec{\Pi}(\vec{x}))^2) - B\sqrt{1 - \vec{\Pi}^2(\vec{x})}]}.$$

$$\tag{5.24}$$

Writing

$$\prod_x \frac{1}{\sqrt{1 - \vec{\Pi}^2(x)}} = e^{-\frac{1}{2a^d} \int d\vec{x} \ln(1 - \vec{\Pi}^2(\vec{x}))}, \tag{5.25}$$

where $a \sim \Lambda^{-1}$ may be understood as the lattice spacing, and then expanding the action to quartic order in $\vec{\Pi}(\vec{x})$, the action becomes

$$S = \frac{1}{2T} \int d\vec{x} \left[(\nabla \vec{\Pi}(\vec{x}))^2 + \frac{1}{4} (\nabla \vec{\Pi}^2(\vec{x}))^2 - \frac{T}{a^d} \vec{\Pi}^2(\vec{x}) \right.$$
$$\left. - \frac{T}{2a^d} \vec{\Pi}^4(\vec{x}) + B \left(\vec{\Pi}^2(\vec{x}) + \frac{1}{4} \vec{\Pi}^4(\vec{x}) \right) + O(\vec{\Pi}^6) \right]. \tag{5.26}$$

Written in this form, the action becomes explicitly non-Gaussian, as there are now quartic terms of three different types. Let us first take the zeroth order approximation and neglect all the quartic terms. At low temperatures we can also neglect the quadratic term that arises from the measure of the integral and write the quadratic part as

$$S_0 = \frac{1}{2T} \int_0^\Lambda \frac{d\vec{k}}{(2\pi)^d} (k^2 + B) \Pi_\alpha(\vec{k}) \Pi_\alpha(-\vec{k}), \tag{5.27}$$

with the sum over $\alpha = 1, 2, \ldots, N - 1$ being assumed. After the usual integration over the fast modes and the change $bk \to k$ to bring the cutoff back to its original value, the magnetic field needs to be rescaled as $b^2 B \to B$. At the level of Gaussian approximation it appears that there is no need to rescale $\Pi_\alpha(k)$. However, before the constraint was resolved, the term involving the magnetic field in the action in Eq. (5.23) was $\sim B\sigma(k = 0)/T$. Since $\sigma(k)$ scales the same as $\Pi_\alpha(k)$, and the action does not scale, the combination B/T and $\Pi_\alpha(k)$ have to be rescaled in exactly opposite ways: $Z_\Pi \vec{\Pi}(\vec{k}) \to \vec{\Pi}(\vec{k})$, $Z_\Pi^{-1} B/T \to B/T$. Comparing with the form in Eq. (5.27) then determines the rescaling factor Z_Π: $Z_\Pi = b^{-d} + O(T)$. As a result the temperature after the integration over the fast modes in the zeroth order approximation becomes changed as

$$T \to T(b) = b^{2+d} Z_\Pi^2 T = b^{2-d} T + O(T^2). \tag{5.28}$$

For $d < 2$, therefore, low temperature is relevant and grows under renormalization. As increase of temperature under renormalization is always expected at high temperatures, for $d < 2$ the ordered phase at $T = 0$ appears unstable and renormalization is always towards the high-temperature region (see Fig. 5.1). For $d > 2$, on the other hand, the $T = 0$ fixed point becomes stable. Since a high temperature should still increase under renormalization there

$T = 0$ $d \le 2$ $T = \infty$

Figure 5.1 The renormalization group flow of the temperature in the non-linear σ-model below lower critical dimension.

should be an unstable fixed point at a finite temperature that separates the two regimes. One thus expects that

$$\beta_T = \frac{\mathrm{d}T}{\mathrm{d}\ln(b)} = (2 - d)T + O(T^2), \qquad (5.29)$$

with a positive $O(T^2)$ term. If this is indeed the case there will be a critical point at

$$T^* \propto (d - 2), \qquad (5.30)$$

which is small close to and above two dimensions. This reasoning therefore supplies us with an alternative perturbation parameter, $\epsilon = d - d_{\mathrm{low}}$. As discussed in the next section, the neglected quartic terms contribute the terms $O(T^n)$ with $n \ge 2$ in β_T. Critical exponents may therefore in principle be obtained from low-temperature expansion close to two dimensions.

5.4 Low-temperature expansion

Consider the thermal average

$$\langle \Pi_\alpha(\vec{k}_1)\Pi_\beta(\vec{k}_2) \rangle = (2\pi)^d G(k_1)\delta_{\alpha\beta}\delta(\vec{k}_1 + \vec{k}_2), \qquad (5.31)$$

with $G^{-1}(k) = (k^2 + B)/T$ to the leading (zeroth) order in T, as follows from Eq. (5.27). The remaining quadratic and the higher-order terms in the action provide the corrections to this form which are of higher order in temperature, and which can be summed into redefinitions of the magnetic field, temperature, and the fields. Since $G(k) \sim T$, to the first order in temperature we can neglect all but the quartic terms in the action, because higher-order terms introduce higher powers in $G(k)$ into the calculation and thus will lead to higher-order corrections in temperature. To the first order in temperature we may also neglect the quartic term that arises from the measure. The two remaining quartic terms in the action (5.26) can be written in the Fourier space as

$$\frac{B}{8T} \int \frac{\mathrm{d}\vec{k}_1 \ldots \mathrm{d}\vec{k}_4}{(2\pi)^{3d}} \delta\left(\sum_{i=1}^{4} \vec{k}_i\right) \Pi_\alpha(\vec{k}_1)\Pi_\alpha(\vec{k}_2)\Pi_\beta(\vec{k}_3)\Pi_\beta(\vec{k}_4), \qquad (5.32)$$

and

$$-\frac{1}{8T} \int \frac{d\vec{k}_1...d\vec{k}_4}{(2\pi)^{3d}} \delta\left(\sum_{i=1}^{4}\vec{k}_i\right)(\vec{k}_1 + \vec{k}_2)\cdot(\vec{k}_3 + \vec{k}_4)\Pi_\alpha(\vec{k}_1)\Pi_\alpha(\vec{k}_2)\Pi_\beta(\vec{k}_3)\Pi_\beta(\vec{k}_4),$$

(5.33)

where summation over the repeated indices is assumed. To the first order in T the integration over the fast modes leads to

$$G^{-1}(q) = \frac{1}{T}(q^2 + B) - \frac{1}{a^d} + (N+1)\frac{B}{2}\int_{\Lambda/b}^{\Lambda} \frac{d\vec{k}}{(2\pi)^d} \frac{1}{k^2 + B}$$

$$+ \int_{\Lambda/b}^{\Lambda} \frac{d\vec{k}}{(2\pi)^d} \frac{q^2 + k^2}{k^2 + B},$$

(5.34)

which can be rewritten as

$$G^{-1}(q) = \frac{q^2}{T}\left(1 + T\frac{\Lambda^{d-2}S_d}{(2\pi)^d}\ln(b) + O(BT, T^2)\right)$$

$$+ \frac{B}{T}\left(1 + T\frac{N-1}{2}\frac{\Lambda^{d-2}S_d}{(2\pi)^d}\ln(b) + O(BT, T^2)\right)$$

$$- \frac{1}{a^d}\left(1 - (\Lambda a)^d\frac{S_d}{(2\pi)^d}\ln(b)\right).$$

(5.35)

The last term implies that the lattice constant has effectively increased after the integration over the fast modes. The action for the remaining slow modes can then be written as

$$S_< = \frac{1}{2T}\int_0^{\Lambda/b} \frac{d\vec{k}}{(2\pi)^d}G^{-1}(k)\Pi_\alpha(\vec{k})\Pi_\alpha(-\vec{k}).$$

(5.36)

Rescaling the field as $Z_\Pi\vec{\Pi}(k) \to \vec{\Pi}(k)$ and $Z_\Pi^{-1}B/T \to B/T$, the above form implies that

$$Z_\Pi = b^{-d}\left(1 + T\frac{N-1}{2}\frac{\Lambda^{d-2}S_d}{(2\pi)^d}\ln(b) + O(T^2)\right).$$

(5.37)

The renormalized temperature then becomes

$$T(b) = b^{2+d}Z_\Pi^2 T\left(1 - T\frac{\Lambda^{d-2}S_d}{(2\pi)^d}\ln(b) + O(BT, T^2)\right)$$

$$= b^{2-d}T\left(1 + T(N-2)\frac{\Lambda^{d-2}S_d}{(2\pi)^d}\ln(b)\right).$$

(5.38)

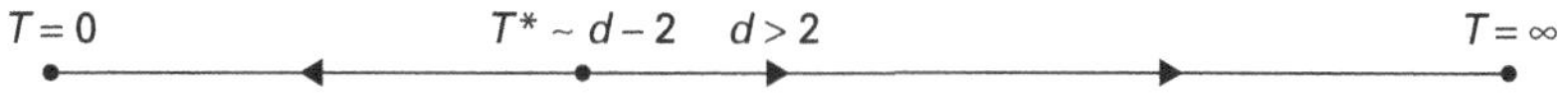

$T = 0$ $T^* \sim d - 2$ $d > 2$ $T = \infty$

Figure 5.2 The flow of temperature in the non-linear σ-model above the lower critical dimension.

Defining the dimensionless temperature as $\hat{T} = T\Lambda^{d-2}S_d/(2\pi)^d$, we may finally write

$$\beta_T = \frac{d\hat{T}}{d\ln(b)} = (2 - d)\hat{T} + (N - 2)\hat{T}^2 + O(\hat{T}^3). \qquad (5.39)$$

If $d > 2$ and $N > 2$ there is indeed a critical point at $\hat{T}^* = (d - 2)/(N - 2) + O((d - 2)^2)$ (Figure 5.2). The slope of the β-function at this critical point determines the correlation length exponent

$$\nu^{-1} = \frac{d\beta_T}{d\hat{T}}\Big|_{\hat{T}=\hat{T}^*} = d - 2 + O((d - 2)^2). \qquad (5.40)$$

From Eqs. (5.35) and (3.44) the anomalous dimension is found to be

$$\eta = \hat{T}^* + O((\hat{T}^*)^2) = \frac{d - 2}{N - 2} + O((d - 2)^2). \qquad (5.41)$$

The remaining exponents follow from the usual scaling laws. For $d \leq 2$ and $N > 2$ the renormalization transformation always increases the temperature, and the only stable fixed point is at $T = \infty$. The interpretation of this flow is that the system is in the disordered phase at any finite temperature.

5.5 Discussion

The above expansion of the exponents in powers of $d - 2$ has been computed to higher orders, but, as may be already suspected from the leading terms, the series does not lead to nearly such accurate values as the expansion around the upper critical dimension. For example, order by order in the $d - 2$ expansion, in the first three orders and for $d = 3$ and $N = 3$ one finds $\nu^{-1} = (1, 2, 2.5)$. Probably the best value extracted from this series and the next, fourth order term, by the Borel–Padé transform, gives $\nu = 0.799$, still much larger than the correct value of $\nu = 0.705$. For the anomalous dimension the situation is even worse, and one finds $\eta = (1, -1, 2)$ in the first three orders.

The reason for this apparent failure of the expansion around the lower critical dimension is not entirely clear at the time of writing. This issue is not only of academic interest, since there are physical problems which may be formulated in terms of an effective non-linear σ-model, and for which a finite upper critical dimension is believed not to exist. A notable example of this is the quantum mechanical wave-function localization in a random potential. The large anomalous dimension seems most worrisome in this respect, because it may be indicating a possible failure of the gradient expansion used to discard the higher-order derivatives of the fields in Eq. (5.23).

For the superfluid $N = 2$, and the quadratic term in β_T vanishes. In fact the same happens with all the higher-order terms as well, and for $N = 2$ and $d = 2$, $\beta_T = 0$ exactly. That this must be so may be seen by realizing that in this case the non-linear σ-model may be written in terms of the phase of the order parameter as

$$S = \frac{|\Phi_0|^2}{2T} \int d\vec{x} (\nabla \phi(\vec{x}))^2, \tag{5.42}$$

and without any constraint on the angle ϕ. Perturbation theory in T is therefore trivial. This however does not mean that there cannot be a phase transition in a two-dimensional superfluid. Perturbation theory considers only small gradients of the phase for which the continuum approximation suffices; however, the continuum expression in Eq. (5.42) is oblivious to the fact that ϕ is actually a periodic variable. The periodicity of the phase leads to a peculiar transition in the two-dimensional superfluid between the algebraically ordered phase at low temperatures and the disordered high-temperature phase, to which we turn in the next chapter.

Problem 5.1 Find the temperature at which the uniform susceptibility would diverge in a two dimensional superfluid in the Gaussian approximation.

Solution The uniform ($k = 0$) superfluid susceptibility is

$$\chi(k = 0) = \int d\vec{x} \langle \Phi^*(\vec{x}) \Phi(0) \rangle.$$

From Eq. (5.5), the above integral diverges if $T/(4\pi |\Phi_0|^2) < 2$.

Problem 5.2 Find the correlation length of the two-dimensional Heisenberg model at low temperatures.

Solution Integrating the flow equation for the temperature, one finds

$$\xi(T) \sim e^{\frac{2\pi \Phi_0^2}{T}},$$

where we restored the amplitude of the order parameter. The correlation length is therefore always finite, but rapidly becomes very long at low temperatures.

Problem 5.3 Consider the anisotropic non-linear σ-model with the action

$$S = \frac{1}{2T} \int d\vec{x}(u(\nabla \sigma(\vec{x}))^2 + (\nabla \vec{\Pi}(\vec{x}))^2),$$

and with $u \neq 1$. Determine the relevance of such weak anisotropy near two dimensions.

Solution Let us assume $u = 1 + \delta$, with $|\delta| \ll 1$, and the magnetic field along the Nth direction. At the magnetic field $B = 0$ and to the linear order in temperature, the action for the slow modes becomes

$$S_< = \frac{1}{2T} \int_0^{\Lambda/b} \frac{d\vec{k}}{(2\pi)^d} \left\{ \left[k^2(1+\delta) + k^2 \frac{\hat{T}}{1+\delta} \ln(b) \right] \sigma(\vec{k})\sigma(-\vec{k}) \right.$$

$$\left. + k^2(1 + \hat{T} \ln(b)) \sum_{\alpha=2}^{N-1} \Pi_\alpha(\vec{k})\Pi_\alpha(-\vec{k}) \right\}.$$

For $\delta \ll 1$, this yields

$$\frac{d\delta}{d \ln(b)} = -2\hat{T}\delta + O(\hat{T}\delta^2),$$

so that weak anisotropy of either sign is irrelevant near $d = 2$.

Problem 5.4* Determine the flow diagram of the non-linear σ-model

$$S = \frac{1}{2T} \int d\vec{x}((\nabla \sigma(\vec{x}))^2 + (\nabla \vec{\Pi}(\vec{x}))^2 + g\vec{\Pi}^2(\vec{x})),$$

where the coupling g explicitly breaks the $O(N)$ symmetry down to $O(N-1) \times Z_2$. For $d = 2$ and $N = 3$ the model may be used to describe the solid–superfluid transition in a system of hard-core bosons on a quadratic lattice. What does the line $g = 0$, $T < T^*$ represent?

Solution First, consider the case $g > 0$. Defining the self-energy as $\Sigma(p) = G^{-1}(p) - (p^2 + B + g)/T$, after the integration over the fast modes one finds

$$\Sigma(p) = -a^{-d} + \frac{B(N+1)}{2} \int_{\Lambda/b}^{\Lambda} \frac{d\vec{k}}{(2\pi)^d} \frac{1}{k^2 + B + g}$$

$$+ \int_{\Lambda/b}^{\Lambda} \frac{d\vec{k}}{(2\pi)^d} \frac{p^2 + k^2}{k^2 + B + g},$$

similarly to Eq. (5.34). Rearranging the terms,

$$\Sigma(p) = \frac{B(N-1)\ln(b)}{4\pi(1+\hat{g})} + p^2 \frac{\ln(b)}{2\pi(1+\hat{g})} - g\frac{\ln(b)}{2\pi(1+\hat{g})},$$

where we neglected the term from the measure and assumed a dimension near two. The fields therefore need to be rescaled as $Z\Pi_\alpha \to \Pi_\alpha$, with

$$Z = b^{-d}\left(1 + T\frac{N-1}{2}\frac{\ln(b)}{2\pi(1+\hat{g})}\right),$$

which implies

$$\frac{d\hat{T}}{d\ln(b)} = (2-d)\hat{T} + \frac{N-2}{(1+\hat{g})}\hat{T}^2,$$

$$\frac{d\hat{g}}{d\ln(b)} = 2\hat{g}\left(1 - \frac{\hat{T}}{1+\hat{g}}\right),$$

where the dimensionless $\hat{g} = g\Lambda^{-2}$, and $\hat{T} = T/(2\pi)$.

For $g < 0$, first we write $g\vec{\Pi}^2 = g + |g|\sigma^2$, and drop the constant term in the action. Then we assume that the field is along the direction of Π_N. We may then write the self-energy for the σ component:

$$\Sigma_\sigma(p) = -a^{-d} + \frac{B}{2}\left((N-2)\int_{\Lambda/b}^{\Lambda} \frac{d\vec{k}}{(2\pi)^d} \frac{1}{k^2 + B}\right.$$

$$\left.+ 3\int_{\Lambda/b}^{\Lambda} \frac{d\vec{k}}{(2\pi)^d} \frac{1}{k^2 + B + |g|}\right) + \int_{\Lambda/b}^{\Lambda} \frac{d\vec{k}}{(2\pi)^d} \frac{p^2 + k^2}{k^2 + B + |g|}$$

$$= \frac{B}{2}\left[\frac{(N-2)\ln(b)}{2\pi} + \frac{\ln(b)}{2\pi(1+|\hat{g}|)}\right]$$

$$+ p^2 \frac{\ln(b)}{2\pi(1+|\hat{g}|)} - |g|\frac{\ln(b)}{2\pi(1+|\hat{g}|)}.$$

Similarly, for $N - 2$ components of $\vec{\Pi}$,

$$\Sigma_\Pi(p) = -a^{-d} + \frac{B}{2}\left(N\int_{\Lambda/b}^{\Lambda}\frac{d\vec{k}}{(2\pi)^d}\frac{1}{k^2 + B} + \int_{\Lambda/b}^{\Lambda}\frac{d\vec{k}}{(2\pi)^d}\frac{1}{k^2 + B + |g|}\right)$$

$$+ \int_{\Lambda/b}^{\Lambda}\frac{d\vec{k}}{(2\pi)^d}\frac{p^2 + k^2}{k^2 + B}$$

$$= \frac{B}{2}\left[\frac{(N-2)\ln(b)}{2\pi} + \frac{\ln(b)}{2\pi(1 + |\hat{g}|)}\right] + p^2\frac{\ln(b)}{2\pi}.$$

The integration over the fast modes therefore changes the couplings as

$$\frac{B}{T} \to \frac{B}{T}\left(1 + \frac{T\ln(b)}{4\pi}\left(N - 2 + \frac{1}{1 + |\hat{g}|}\right)\right),$$

and

$$\frac{g}{T} \to \frac{g}{T}\left(1 - \frac{T\ln(b)}{2\pi(1 + |\hat{g}|)}\right),$$

before the momenta are rescaled to restore the cutoff. We then rescale $Z_\Pi\Pi_\alpha \to \Pi_\alpha$, with

$$Z_\Pi = b^{-d}\left(1 + \frac{T\ln(b)}{4\pi}\left(N - 2 + \frac{1}{1 + |\hat{g}|}\right)\right).$$

The rescaled temperature is therefore

$$T(b) = b^{2+d}Z_\Pi^2 T\left(1 - \frac{T}{2\pi}\ln(b)\right),$$

which yields

$$\frac{d\hat{T}}{d\ln(b)} = (2 - d)\hat{T} + \left(N - 3 + \frac{1}{1 + |\hat{g}|}\right)\hat{T}^2.$$

Since the $\sim p^2$ terms in Σ_Π and Σ_σ are different, the field σ needs to be rescaled differently to be in accord with the obtained renormalization of temperature: $Z_\sigma\sigma \to \sigma$, with $Z_\sigma \neq Z_\Pi$. Since one may also write

$$T(b) = b^{2+d}Z_\sigma^2 T\left(1 - \frac{T}{2\pi(1 + |g|)}\ln(b)\right),$$

the comparison between the two forms for $T(b)$ gives

$$Z_\sigma = b^{-d}\left(1 + \frac{T\ln(b)}{4\pi}\left(N - 2 + \frac{1 - |g|}{1 + |g|}\right)\right).$$

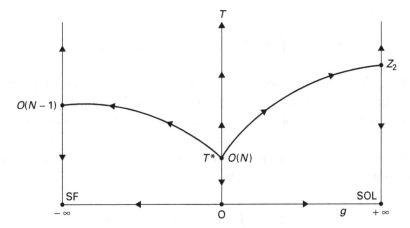

Figure 5.3 The flow of the symmetry breaking parameter g and the temperature T. There is a first-order transition between the superfluid and the solid phase below T^*.

Finally, since

$$\frac{g(b)}{T(b)} = \frac{g}{T} b^{-d} Z_\sigma^{-2} \left(1 - \frac{T \ln(b)}{2\pi (1 + |g|)} \right),$$

one obtains

$$\frac{d\hat{g}}{d \ln(b)} = 2\hat{g} \left(1 - \frac{\hat{T}}{1 + |\hat{g}|} \right),$$

for $g < 0$.

The flow diagram in $d > 2$ is depicted schematically in Fig. 5.3. The $O(N)$-symmetric point at $g = 0$ is unstable in both the T and g directions, and is thus bicritical. For $g > 0$ the flow is towards the Ising critical point at $\hat{g} = \infty$ and $\hat{T}^* = \infty$, the latter infinity serving as a warning that the discrete Ising symmetry does not allow any Goldstone modes, and the σ-model description is inadequate. For $g < 0$, on the other hand, the flow is towards the $O(N-1)$ critical point at $\hat{g} = -\infty$ and $\hat{T}^* = (d-2)/(N-3)$. For $T < T^*$ and $g = 0$ the flow is towards the fixed point at $T = g = 0$, which describes a first-order transition. For $N = 3$ this corresponds to the discontinuous superfluid–solid transition for lattice bosons.

Problem 5.5* Consider the d-dimensional $O(N)$-symmetric Ginzburg–Landau–Wilson theory with an additional long-range interaction between the fields of the form

$$S' = \frac{1}{4T} \int d\vec{x}\, d\vec{y}\, [\Phi_i(\vec{x}) - \Phi_i(\vec{y})]^2 \frac{V}{|\vec{x} - \vec{y}|^{d+r}}.$$

Such a term would represent a long-range interaction between the dipoles in the Ising, the XY, and the Heisenberg models, when $N = 1, 2, 3$, respectively. Determine the stability of the ordered phase as a function of the power r.

Solution When rewritten in the Fourier space the above addition to the Ginzburg–Landau–Wilson action becomes

$$S' = \frac{C(d,r)}{2T} \int \frac{d\vec{k}}{(2\pi)^d} \Phi_i(\vec{q}) |\vec{q}|^r \Phi_i(-\vec{q}),$$

with the coefficient

$$C(d,r) = V \int d\vec{x} \frac{1 - e^{ix_d}}{|\vec{x}|^{d+r}}$$

finite for $0 < r < 2$. For this range of r the original $\sim q^2$ term in the quadratic part of the action becomes irrelevant compared to the new $\sim |\vec{q}|^r$ term, and can be neglected. Repeating the analysis of the resulting non-linear σ-model with q^2 replaced by $|\vec{q}|^r$ to the zeroth order in temperature yields

$$T(b) = b^{r+d} Z_\Pi^2 T + O(T^2)$$

with $Z_\Pi = b^{-d}(1 + O(T))$, and therefore

$$\beta_T = (r - d)T + O(T^2).$$

So for $d > 2$, $r - d < 0$, and the ordered phase is always stable, as it was for the usual short-range interaction alone. For $d < 2$, the ordered phase becomes stable if $r < d$, and remains unstable otherwise. Long-ranged interactions, if sufficiently slowly decaying with the distance, have therefore a stabilizing effect on the ordered phase.

Problem 5.6* Determine the correlation length exponent at the ferromagnetic transition in the one-dimensional Heisenberg model with interactions $\sim |\vec{x}|^{-19/10}$ between dipoles.

Solution From the previous problem it follows that for $d = 1$ and $r = 9/10$ a low finite temperature is irrelevant, and the ordered ferromagnetic phase is stable. To determine the properties at the transition we need the higher-order terms in β_T. The calculation is completely equivalent to the one in Section 5.4, with one important caveat. Since the leading term in the quadratic action for $\vec{\Pi}$ modes is non-analytic in q^2 when $r < 2$, it cannot get generated by the momentum-shell integration. This implies that

$$T(b) = b^{r+d} Z_\Pi^2 T,$$

with $Z_\Pi = b^{-d}(1 + cT(N-1)\ln(b) + (T^2))$, with c as a cutoff-dependent constant, similarly as in Eq. (5.37). Redefining the temperature as $cT \to T$ we thus find

$$\beta_T = (r-d)T + (N-1)T^2 + O(T^3).$$

Note the change in the coefficient of the T^2 term as compared to Eq. (5.39). The correlation length exponent is therefore

$$\nu^{-1} = d - r + O((d-r)^2),$$

which for $d = 1$ and $r = 9/10$ gives $\nu \approx 10$.

6

Kosterlitz–Thouless transition

Vortex configurations of the phase of the superfluid order parameter are introduced. Unbinding of vortex–antivortex pairs at the transition is described first in a mean-field theory for the two-dimensional Coulomb gas. The XY model is shown to be related by duality to the sine-Gordon theory. The latter is then studied systematically by the renormalization group at low vortex fugacity, and the critical behavior at the superfluid transition in two dimensions is determined in this way. Finally, some important differences in topology between the XY and the Heisenberg model are pointed out, and the reasons for the absence of a Kosterlitz–Thouless transition in the latter are discussed.

6.1 Vortices and spin waves

The phase of the order parameter is a periodic variable and this has profound consequences for the superfluid order, particularly in two dimensions. Consider, for example, the configuration of the phase depicted in Fig. 6.1 (a), where we have placed the order parameter on a discrete lattice. As one encircles the central plaquette counter clockwise the phase represented by arrows changes by 2π. This is true for any contour irrespectively of its size, as long as it encircles the central plaquette. Such a configuration will be called a *vortex*. Note that, although far from the center of the vortex the deviation of the phase between neighboring lattice sites is small, this is not so close to the center where the phase has to change by 2π over a small distance. For such a configuration the assumption that $\nabla\phi$ is everywhere small manifestly fails, and the low-temperature expansion from the previous chapter becomes inadequate.

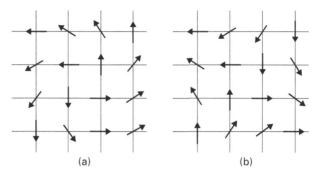

Figure 6.1 Vortex (a) and antivortex (b) on a quadratic lattice.

An arbitrary configuration of the phase in continuum may be divided into parts with and without vortices when written as

$$\phi(\vec{x}) = \psi(\vec{x}) + \theta(\vec{x}), \tag{6.1}$$

where over any closed contour

$$\oint d\vec{x} \cdot \nabla\psi(\vec{x}) = 0, \tag{6.2}$$

and

$$\oint d\vec{x} \cdot \nabla\theta(\vec{x}) = 2\pi q. \tag{6.3}$$

q is a positive or negative integer. $\psi(\vec{x})$ represents the smooth part of the phase, often called the *spin wave*, and $\theta(\vec{x})$ represents the vortex part. The *antivortex* with $q = -1$ is depicted in Fig. 6.1 (b).

One can easily estimate the energy of an isolated vortex. Assuming the center of the vortex to be at $\vec{x} = 0$, the configuration with minimum energy for a vorticity q is

$$\theta(\vec{x}) = q\alpha, \tag{6.4}$$

with α being the polar angle between $\vec{x}$ and an arbitrary fixed axis (Problem 6.1). Then $\nabla\theta(\vec{x}) = (q/|\vec{x}|)\hat{\alpha}$, and the energy of the vortex in a system of a finite linear size R and in the continuum approximation is

$$H_v = \frac{\Phi_0^2}{2} \int d\vec{x}(\nabla\theta(\vec{x}))^2 = \pi q^2 \Phi_0^2 \int_{r_0}^{R} \frac{r\,dr}{r^2} = \pi q^2 \Phi_0^2 \ln\frac{R}{r_0}. \tag{6.5}$$

r_0 is a short-distance cutoff, which may be understood as proportional to the lattice spacing. Evidently, r_0 needs to be sufficiently larger than the lattice spacing for the continuum approximation to vortex energy to be sensible. The result in Eq. (6.5) excludes therefore the finite part of the vortex energy which comes from the vicinity of the center of the vortex, and is often referred to as the *core energy*. It shows, however, that in the thermodynamic limit $R \to \infty$ the energy of an isolated vortex or an antivortex diverges. One may therefore expect that at low temperatures vortices are indeed negligible in a large system. Their presence, however, increases the entropy of the system. The Boltzmann entropy of a single vortex is $\Delta S_v = k_B T \ln(R/r_0)^2$, since the center of a vortex may be placed at any of $\sim (R/r_0)^2$ sites. The increase of the free energy due to an isolated vortex is therefore:

$$\Delta F = H_v - T \Delta S_v = \pi q^2 \Phi_0^2 \ln \frac{R}{r_0} - 2T \ln \frac{R}{r_0}, \tag{6.6}$$

with the Boltzmann constant hereafter set to $k_B = 1$. As both the energy and the entropy depend logarithmically on the system size, ΔF changes sign at

$$T_{KT} = \frac{\pi \Phi_0^2}{2}, \tag{6.7}$$

where we have assumed $q^2 = 1$, which gives the lowest energy. For $T > T_{KT}$, the above argument suggests that vortices become favored enough by the entropy term in the free energy to become free, while for $T < T_{KT}$ they indeed are unimportant due to their prohibitively large energy. Assuming that there is a single phase transition between the disordered high-temperature and the algebraically ordered low-temperature phase, we may tentatively identify it with the appearance of free vortices.

Although this simple argument will indeed turn out to be correct, it certainly is not obviously so. Even below the *Kosterlitz–Thouless* transition temperature T_{KT} thermal fluctuations will inevitably create an equal number of vortices and antivortices in the system. The energy of a vortex–antivortex pair is finite and at large separations equal to

$$H_{\text{pair}} = 2\pi q^2 \Phi_0^2 \ln \frac{|\vec{x}_1 - \vec{x}_2|}{r_0}. \tag{6.8}$$

The expression for H_{pair} follows uniquely from the facts that: (1) at distances much larger than the pair's size there is no net vorticity, so the energy of the pair must be finite, and (2) as the pair's size diverges H_{pair} should yield the sum of energies for an isolated vortex and antivortex. As T_{KT} is approached from below the interaction between the vortex and the antivortex in the largest pair will thus be screened by numerous thermally induced smaller pairs. It is not clear a priori what the effect of a finite density of such vortex–antivortex pairs will be. We turn to this issue in the next section.

6.2 Mean-field theory

For a given distribution of vortices the energy in the continuum approximation may be written as

$$H = \frac{\Phi_0^2}{2} \int d\vec{x} \left[(\nabla \theta(\vec{x}))^2 + (\nabla \psi(\vec{x}))^2 \right], \qquad (6.9)$$

where $\theta(\vec{x})$ is chosen to be a local minimum of H, and $\psi(\vec{x})$ is the (spin-wave) deviation from it. Since $\theta(\vec{x})$ is the local minimum the cross-term $\nabla \psi \cdot \nabla \theta$ vanishes. This means that spin-wave and vortex configurations may be considered decoupled, and the partition function factorizes into the product of the two. Since the spin-wave part leads only to an analytic contribution to the free energy, the only singularity can come from the vortex part. This, on the other hand, can be studied independently from the spin-waves, which we therefore drop in the rest of this section.

At low temperatures we may further assume a low density of bound vortex–antivortex pairs of unit vorticity, so that the vortex partition function in the grand-canonical ensemble may be written as a product over single-pair partition functions:

$$Z_v = \sum_{p=0}^{\infty} \frac{y^{2p} Z_{\text{pair}}^p}{p!}, \qquad (6.10)$$

where $y = e^{\mu/T}$ is the vortex fugacity. The chemical potential measures the finite core energy of the vortex. We will take μ to be negative and large enough so that fugacity may be assumed to be small at temperatures at which we expect bound pairs. The single-pair partition function is

$$Z_{\text{pair}} = \int \frac{d\vec{x}_1 d\vec{x}_2}{r_0^4} e^{-\frac{V(|\vec{x}_1 - \vec{x}_2|)}{T}}. \qquad (6.11)$$

The interaction is $V(r) = (1/2\pi \epsilon_0) \ln(r/r_0)$, with $\epsilon_0 = 1/4\pi^2 \Phi_0^2$. The interactions between vortices belonging to different pairs as well as vortices of higher vorticity are neglected in the low-density approximation.

The partition function Z_v may be straightforwardly computed:

$$\ln Z_v = y^2 Z_{\text{pair}} = \frac{2\pi y^2 A}{r_0^4} \int_{r_0}^{\infty} r \, dr \, e^{-\frac{V(r)}{T}}, \qquad (6.12)$$

where A is the area of the system. The total density of pairs is therefore

$$n = \frac{T}{2A} \frac{d \ln Z_v}{d\mu} = \int_{r_0}^{\infty} dn(r), \qquad (6.13)$$

where $dn(r)$ is the density of pairs with the size between r and $r + dr$:

$$dn(r) = \frac{2\pi y^2}{r_0^4} e^{-\frac{V(r)}{T}} r\, dr. \tag{6.14}$$

We may account for the screening effects of smaller pairs on the interaction between the vortices of a larger pair by including a dielectric constant in $V(r)$ for the large pair. Since when $r \approx r_0$ there are no smaller pairs, the dielectric constant at small distances should be close to unity, whereas as $r \to \infty$ it should approach its macroscopic value. This suggests an introduction of an r-dependent dielectric constant $\epsilon(r)$, where

$$dV_{\text{eff}}(r) = \frac{dV(r)}{\epsilon(r)}. \tag{6.15}$$

Assuming $\epsilon(r)$ to be a slowly varying function, we can rewrite the above equation as

$$V_{\text{eff}}(r) \approx \frac{V(r)}{\epsilon(r)}, \tag{6.16}$$

and then replace $V(r)$ with $V_{\text{eff}}(r)$ in Eq. (6.14). This assumption will be justified a posteriori near the transition and in the dilute limit.

The idea of the mean-field theory is to determine the thermodynamic value of the dielectric constant $\epsilon(r \to \infty)$, and in particular to check if it may become divergent as the temperature is changed. In the language of electrostatics, such a divergence would signal a phase transition from the "dielectric" phase of bound dipoles into the "metallic" phase of free charges with perfect screening. One thus needs to relate $n(r)$ to $\epsilon(r)$ to obtain the equation for the latter. This may be done by computing the "polarizability" of the system as follows. Considering vortex–antivortex pairs as electrical dipoles, the energy of a single dipole of fixed size r in a fictitious "electric field" E will be

$$H(E) = -Er \cos\alpha, \tag{6.17}$$

so that the polarizability of such a dipole is

$$\chi(r) = \frac{d}{dE} \int_0^{2\pi} \frac{d\alpha}{2\pi} (r \cos\alpha) e^{-H(E)/T}|_{E=0} = \frac{1}{2T} r^2. \tag{6.18}$$

The dielectric constant at the distance r is therefore

$$\epsilon(r) = 1 + \frac{1}{\epsilon_0} \int_{r_0}^{r} \chi(r') dn(r'), \tag{6.19}$$

or, recast in differential form,

$$d\epsilon(r) = \frac{\pi y^2}{T\epsilon_0} \left(\frac{r}{r_0}\right)^{4-\frac{2\pi\Phi_0^2}{T\epsilon(r)}} d\ln\frac{r}{r_0}. \tag{6.20}$$

Equation (6.20) is the desired differential equation for the r-dependent dielectric constant. Let us rewrite it in a more convenient form by introducing the dimensionless quantities $x = \ln(r/r_0)$ and $z = 2\pi T\epsilon_0\epsilon(r)$:

$$\frac{dz}{dx} = 2\pi^2 y^2 e^{-x(\frac{1}{z}-4)}. \tag{6.21}$$

We are interested in the asymptotic value of z as $x \to \infty$, subject to the initial condition

$$z(0) = \frac{T}{2\pi\Phi_0^2}, \tag{6.22}$$

that is $\epsilon(r_0) = 1$.

At low temperatures $z(0) \ll 1$, and the exponent in Eq. (6.21) is large and negative. This means that z increases slowly and may saturate at a finite value. Indeed, for $y \ll 1$ and finite $z(\infty) \ll 1$ we find $z(\infty) = z(0) + O(y^2)$, and to the leading order in fugacity we may replace z with $z(0)$ in the exponent. This gives

$$z(x) = z(0)\left[1 + \frac{2\pi^2 y^2}{1 - 4z(0)}\left(1 - e^{-x(\frac{1}{z(0)}-4)}\right) + O(y^4)\right]. \tag{6.23}$$

The highest temperature for which $z(\infty)$ is finite is therefore determined by the condition

$$z(\infty) = \frac{1}{4}. \tag{6.24}$$

The solutions at various temperatures are plotted schematically in Fig. 6.2. At low fugacity we may determine the critical temperature as follows. Define a new variable $u > 0$ as

$$z = \frac{1-u}{4}, \tag{6.25}$$

so that at $T = T_c$, $u(\infty) = 0$, and from Eq. (6.22),

$$\frac{T_c}{2\pi\Phi_0^2} = \frac{1-u(0)}{4}. \tag{6.26}$$

Since at low fugacity $u \ll 1$, we can neglect the terms of higher order in u in the exponent in Eq. (6.21). Upon redefining the variables as $\tilde{u} = u/(\sqrt{2\pi}y)$

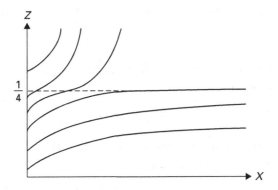

Figure 6.2 Solutions of Eq. (6.21), with $x = \ln(r/r_0)$ and $z = 2\pi T \epsilon_0 \epsilon(r)$.

and $\tilde{x} = 4\sqrt{2\pi} yx$ Eq. (6.21) transforms into

$$\frac{d\tilde{u}}{d\tilde{x}} = -e^{-\tilde{u}\tilde{x}} + O(y). \tag{6.27}$$

Defining $\tilde{u}_c$ as the smallest $\tilde{u}(0)$ for which $\tilde{u}(\infty) \geq 0$ in the last equation, one finally obtains

$$T_c = \frac{\pi \Phi_0^2}{2} \left(1 - \tilde{u}_c \sqrt{2\pi} y + O(y^2) \right). \tag{6.28}$$

The precise value of $\tilde{u}_c$ can be determined numerically, for example; we find $\tilde{u}_c \approx 1.7$.

A finite density of pairs therefore screens the interaction between the vortex and the antivortex in the largest pair and consequently makes it easier to dissociate: $T_c < T_{KT}$ for a finite y. The nature of the transition is not changed when the fugacity is low, however; the transition is still due to unbinding of the vortices in the largest pair, only at a somewhat lower transition temperature. Note that when $y \ll 1$, $\epsilon(r)$ is indeed a slowly varying function below T_c, so the assumption made in Eq. (6.16) was after all justified. At $T > T_c$, $z(\infty) \to \infty$ for any fugacity, and our approximation evidently breaks down. It is natural to identify this phase with the plasma of free vortices which then provide perfect screening.

It is possible to proceed further and determine the critical behavior implied by the above mean-field theory. Since mean-field theories are typically not reliable in the critical region, instead of doing this in the next section we will rewrite the problem in a form that will facilitate a more systematic treatment of the transition and the critical region.

Problem 6.1 Show that the pure vortex in Eq. (6.4) is a local minimum of the action in Eq. (5.42).

Solution Let us assume a general configuration of the phase with the vorticity q centered at the origin:

$$\phi(\vec{r}) = q\alpha + \psi(\vec{r}),$$

where $\psi(\vec{r})$ is a spin-wave deviation from the pure vortex. The energy density in such a configuration is proportional to

$$(\nabla\phi(\vec{r}))^2 = q^2(\nabla\alpha)^2 + (\nabla\psi(\vec{r}))^2 + 2q\nabla\alpha \cdot \nabla\psi(\vec{r}).$$

Since

$$\nabla\alpha = \frac{-y\hat{x} + x\hat{y}}{|\vec{r}|^2} = -\nabla \times (\hat{z}\ln|\vec{r}|)),$$

where $\hat{z}$ is the unit vector orthogonal to the plane, the last term in the energy density vanishes. The action in Eq. (5.42) is minimized therefore by $\psi(\vec{r}) \equiv 0$, which is just the pure vortex in Eq. (6.4).

Problem 6.2* Consider a dilute three-dimensional neutral system of unit charges interacting via a logarithmic interaction. What phases do you expect?

Solution If screening is completely neglected, a single dipole would dissociate at the temperature determined by the energy–entropy argument in Eq. (6.6), adjusted to three dimensions:

$$\Delta F = q^2 \ln\frac{R}{r_0} - 3T_c \ln\frac{R}{r_0} = 0.$$

We assumed $\pi\Phi_0^2 = k_B = 1$ for simplicity. For $q = 1$, $T_c = 1/3$. We may check the stability of the dipole phase in the presence of a finite density of charges as follows. Assume that the distribution of charges, $\rho(\vec{x})$, interacts via a $-\ln|\vec{x}|$ interaction in three dimensions, and is located in a finite region of size R. At a distance $x \gg R$ we can write the interaction as

$$V(\vec{x}) = \int d^3\vec{x}'\rho(\vec{x}')\left(-\ln x + \frac{\vec{x} \cdot \vec{x}'}{x^2} + \cdots\right)$$

in the spirit of multipole expansion. Thus, for a medium with a charge density $\rho(\vec{x})$ and the dipole moment density $\vec{P}(\vec{x}) = \rho(\vec{x})\vec{x}$, the potential is given by

$$V(\vec{x}) = \int d^3\vec{x}'(-\ln|\vec{x} - \vec{x}'|)(\rho(\vec{x}') - \nabla' \cdot \vec{P}(\vec{x}') + \cdots).$$

The energy of a small dipole in a weak potential is still given by $\vec{P} \cdot \nabla V(\vec{x})$. Assuming non-interacting dilute dipoles at a temperature T then gives the average dipole moment at $\vec{x}$:

$$\langle \vec{P}(\vec{x}) \rangle = -\frac{\langle r^2 \rangle}{3T} \nabla V(\vec{x}).$$

Since the density of dipoles is proportional to y^2, with y being the fugacity, the polarizability of the medium is

$$\chi \propto \frac{\langle r^2 \rangle}{3T} y^2,$$

where the constant of proportionality depends on precise way of imposing the short-distance cutoff. Using the above multipole expansion, the interaction due to a point charge Q in presence of finite density of dipoles is therefore

$$V(k) = \frac{Q}{|\vec{k}|^3 + \chi \vec{k}^2},$$

or, in real space,

$$V(\vec{x}) = \frac{1}{4\pi\chi} \frac{1}{|\vec{x}|} + O\left(\frac{1}{\vec{x}^2}\right).$$

In the presence of a finite density of dipoles, the original logarithmic interaction is screened into the Coulomb $\sim 1/x$ potential at large distances. This suggests that the charges in large dipole pairs interact via the Coulomb potential, and thus should be free. So the finite fugacity would be expected to destabilize the dielectric phase of the system in three dimensions and invalidate the conclusion of the simple energy–entropy argument, in stark contrast to two dimensions.

6.3 Duality and the sine-Gordon theory

The lattice model that in the Gaussian approximation and the continuum limit reduces to Eq. (5.42), but which explicitly incorporates the periodicity of the phase, is the XY model:

$$Z_{xy} = \int_0^{2\pi} \prod_i \frac{d\phi_i}{2\pi} e^{\frac{J}{T} \sum_{i,\hat{\mu}} \cos(\phi_i - \phi_{i+\hat{\mu}})}, \tag{6.29}$$

introduced first in Eq. (1.4) and studied in connection to the Mermin–Wagner–Hohenberg theorem in Section 5.2. The index i labels the sites of a two-dimensional quadratic lattice, $\hat{\mu} = \hat{x}, \hat{y}$, and the coupling constant $J = \Phi_0^2$. When expanded to second order, the XY model reduces to Eq. (5.42) in the continuum limit. It also corresponds to the $\mu \to \infty$, $\lambda \to \infty$ limit of the lattice version of Landau's action for single complex field in Eq. (2.36), with the constant ratio μ/λ. The XY model therefore represents the low-temperature limit of the Φ^4-theory with a single complex field, with the fluctuations of the amplitude of the order parameter suppressed, and should have the transition in the same universality class as the neutral superfluid. It also describes a magnetic system where the local magnetization is confined to a plane.

Having the XY model allows one to put the intuitive ideas of the previous section on firmer ground. First, at low temperatures the important configurations are those where the difference of the phases is close to a multiple of 2π. This suggests the replacement of Z_{xy} by its *Villain approximation*

$$Z_{xy} \approx Z_V = \int_0^{2\pi} \left(\prod_i \frac{d\phi_i}{2\pi} \right) \prod_{i,\hat{\mu}} \sum_{m_{i,\hat{\mu}}=-\infty}^{\infty} e^{-\frac{J}{2T} \sum_{i,\hat{\mu}}(\phi_i - \phi_{i+\hat{\mu}} - 2\pi m_{i,\hat{\mu}})^2}, \quad (6.30)$$

where $m_{i,\hat{\mu}}$ is an integer associated with the link between the sites i and $i + \hat{\mu}$. The advantage of Villain's approximation is that the phases can be integrated out exactly. We may rewrite Z_V using the Hubbard–Stratonovich transformation (Appendix A):

$$Z_V = \int_0^{2\pi} \left(\prod_i \frac{d\phi_i}{2\pi} \right) \prod_{i,\hat{\mu}} \sum_{m_{i,\hat{\mu}}=-\infty}^{\infty} \int_{-\infty}^{\infty} \frac{dx_{i,\hat{\mu}}}{\sqrt{\pi T/2J}}$$

$$\times e^{-\frac{T}{2J} \sum_{i,\hat{\mu}} x_{i,\hat{\mu}}^2 - i \sum_{i,\hat{\mu}} x_{i,\hat{\mu}}(\phi_i - \phi_{i+\hat{\mu}} - 2\pi m_{i,\hat{\mu}})}, \quad (6.31)$$

which can also be directly verified by performing the Gaussian integral over the auxiliary variables $\{x_{i,\hat{\mu}}\}$. Using the Poisson resummation formula

$$\sum_{m=-\infty}^{\infty} e^{-i2\pi m x} = \sum_{n=-\infty}^{\infty} \delta(x - n), \quad (6.32)$$

the sum over $\{m_{i,\hat{\mu}}\}$ restricts the auxiliary variables to strictly integer values, and so

$$Z_V = \int_0^{2\pi} \left(\prod_i \frac{d\phi_i}{2\pi} \right) \prod_{i,\hat{\mu}} \sum_{n_{i,\hat{\mu}}=-\infty}^{\infty} e^{-\frac{T}{2J} \sum_{i,\hat{\mu}} n_{i,\hat{\mu}}^2 - i \sum_{i,\hat{\mu}} n_{i,\hat{\mu}}(\phi_i - \phi_{i+\hat{\mu}})}, \quad (6.33)$$

where the unimportant constant multiplicative factor has been omitted. The phases can now be integrated out exactly by noticing that

$$\int_0^{2\pi} \left(\prod_i \frac{d\phi_i}{2\pi} \right) e^{-i\sum_{i,\hat{\mu}} n_{i,\hat{\mu}}(\phi_i - \phi_{i+\hat{\mu}})} = \int_0^{2\pi} \left(\prod_i \frac{d\phi_i}{2\pi} \right) e^{-i\sum_{i,\hat{\mu}} \phi_i(n_{i,\hat{\mu}} - n_{i-\hat{\mu},\hat{\mu}})}$$

$$= \prod_i \delta(\Delta \cdot \vec{n}_i), \quad (6.34)$$

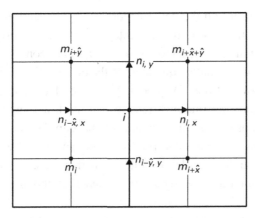

Figure 6.3 The currents $\{\vec{n}_i\}$ on the original lattice and the variables $\{m_i\}$ on the dual lattice.

where Δ denotes a lattice gradient (finite difference):

$$\Delta \cdot \vec{n}_i = \sum_{\hat{\mu}}(n_{i,\hat{\mu}} - n_{i-\hat{\mu},\hat{\mu}}),\tag{6.35}$$

and $\vec{n}_i = (n_{i,\hat{x}}, n_{i,\hat{y}})$. In the Villain approximation the XY model therefore exactly reduces to

$$Z_V = \sum_{\{\vec{n}_i\}}\delta(\Delta \cdot \vec{n}_i)e^{-\frac{T}{2J}\sum_i \vec{n}_i^2}.\tag{6.36}$$

The last expression is also known as the current representation of the XY model, since the constraint in Z_V may be understood as the law of conservation of current.

The constraint in Eq. (6.36) may be resolved by defining another integer variable $\{m_i\}$ at the sites of the dual lattice, formed by the centers of the plaquettes of the original lattice (Fig. 6.3). By writing

$$n_{i,\hat{x}} = m_{i+\hat{x}+\hat{y}} - m_{i+\hat{x}},\tag{6.37}$$

$$n_{i,\hat{y}} = m_{i+\hat{y}} - m_{i+\hat{x}+\hat{y}},\tag{6.38}$$

$$n_{i-\hat{x},\hat{x}} = m_{i+\hat{y}} - m_i,\tag{6.39}$$

$$n_{i-\hat{y},\hat{y}} = m_i - m_{i+\hat{x}},\tag{6.40}$$

we see that

$$\nabla \cdot \vec{n}_i = n_{i,\hat{x}} - n_{i-\hat{x},\hat{x}} + n_{i,\hat{y}} - n_{i-\hat{y},\hat{y}} = 0,\tag{6.41}$$

for any configuration $\{m_i\}$. We may therefore rewrite the partition function as

$$Z_V = \sum_{\{m_i\}}e^{-\frac{T}{2J}\sum_{i,\hat{\mu}}(m_{i+\hat{\mu}}-m_i)^2},\tag{6.42}$$

without any constraint on the new integers $\{m_i\}$. The summation is over the sites of the dual lattice. Note that the integers $\{m_i\}$ are determined uniquely by the configuration of $\{\vec{n}_i\}$, up to an overall additive constant.

The last equation, formally equivalent to the original form of Z_V, represents the *dual* formulation of the XY model. Note two characteristic features: (a) the temperature axis became inverted, so that the low-temperature phase of the XY model maps onto high-temperature phase in the dual model and vice versa, (b) whereas the original variables in the XY model were continuous, the variables in the dual model are integer. If this were not so and the $\{m_i\}$ were real, the dual model would have been Gaussian and thus without a phase transition.

With the help of the Poisson formula, however, we can turn the integers $\{m_i\}$ into continuous real variables by introducing yet another set of integers $\{n_i\}$ as

$$Z_V = \int_{-\infty}^{\infty} \prod_i d\theta_i \sum_{\{n_i\}} e^{-\frac{T}{2J}\sum_{i,\hat{\mu}}(\theta_{i+\hat{\mu}}-\theta_i)^2 - i2\pi\sum_i \theta_i n_i}, \tag{6.43}$$

so that summation over $\{n_i\}$ would bring back Eq. (6.42). To make progress analytically, we now modify the exponent in the last equation by adding a "chemical potential" for the integer variables $\{n_i\}$, i.e. the term

$$\sum_i \ln(y)n_i^2, \tag{6.44}$$

where y is the fugacity. Strictly speaking, $y = 1$ in the Villain model. Generalizing to $y < 1$, however, should not modify Z_V in an essential way. Imagine we integrated over the variables $\{\theta_i\}$ in Eq. (6.43). This would introduce a quadratic term in the action $\sim \sum_{i,j} n_i V_{ij} n_j$, where $V_{ij} \sim \ln |\vec{x}_i - \vec{x}_j|$ for large separations. The integers $\{n_i\}$ therefore represent vortices in the previous section which interact via long-ranged logarithmic interaction. Modifying only the short-range part of such a long-range interaction with the term in Eq. (6.44) should therefore not change the nature of the phase transition in the theory. After all, it is equivalent to a change in the short-distance cutoff r_0 in the interaction energy in Eq. (6.8).

After this generalization Z_V can be simplified by writing

$$\sum_{n=0,\pm 1,\pm 2,\ldots} y^{n^2} e^{-i2\pi\theta n} = 1 + 2y\cos(2\pi\theta) + 2y^4\cos(4\pi\theta) + \cdots$$

$$\approx e^{2y\cos(2\pi\theta)+O(y^2)}. \tag{6.45}$$

For fugacity low enough, only the leading term in the exponent needs to be retained, and

$$Z_V \approx Z_{sG} = \int_{-\infty}^{\infty} \prod_i d\theta_i e^{-\frac{T}{2J}\sum_{i,\mu}(\theta_{i+\mu}-\theta_i)^2 + 2y\sum_i \cos(2\pi\theta_i)}. \tag{6.46}$$

The partition function Z_{sG} defines the *sine-Gordon* model, and, modulo approximations we made in the derivation, is dual to the two-dimensional XY model. In particular, the real variables $\{\theta_i\}$ are dual to the original angles $\{\phi_i\}$ in the following sense. At high temperatures, we see that a gradient of $\{\theta_i\}$ is energetically costly, and $\{\theta_i\}$ should be ordered. At the same high temperatures, on the other hand, the original phases $\{\phi_i\}$ will be disordered. Similarly, at low temperatures $\{\theta_i\}$ are disordered, $\{\phi_i\}$ are ordered. This is why the dual phases $\{\theta_i\}$ are sometimes called *disorder variables*. They provide an alternative point of view at the phase transition in terms of variables which parameterize the degree of disorder for the original degrees of freedom.

Finally, we can write the sine-Gordon theory in the continuum limit as

$$Z_{sG} = \int D\theta(\vec{x}) e^{-\int d\vec{x} L_{sG}[\theta(\vec{x})]}, \tag{6.47}$$

where

$$L_{sG} = \frac{T}{2J}(\nabla\theta(\vec{x}))^2 - 2ya^{-2}\cos(2\pi\theta(\vec{x})), \tag{6.48}$$

where a is the lattice spacing, and with the assumed ultraviolet cutoff on wavevectors of $\Lambda \sim 1/a$. Although we arrived at it through a set of transformations on a lattice, it can also be shown that Eq. (6.47) represents the grand canonical partition function for the system of vortices and antivortices interacting via logarithmic interaction (Problem 6.3). This suggests the identification of the coupling constant y with the fugacity of the single vortex in the XY model. The last expression provides then the basis for the renormalization group analysis to be developed in the next section.

Problem 6.3 Show that the sine-Gordon partition function in Eqs. (6.47)–(6.48) in d dimensions describes a neutral plasma of unit charges interacting via the Coulomb interaction that obeys the Poisson equation $-\nabla^2 V(\vec{x}) = 4\pi^2 \rho(\vec{x})$.

Solution The grand-canonical partition function of a plasma of unit charges in d dimensions is

$$Z = \sum_{n=0}^{\infty} \sum_{m=0}^{\infty} \frac{y_+^n y_-^m}{n! m!} \int d\vec{x}_1 ... d\vec{x}_{n+m} e^{-\frac{1}{2T} \sum_{i,j=1}^{n+m} q_i q_j V(\vec{x}_i - \vec{x}_j)},$$

where $q_i = \pm 1$, and y_+ (y_-) is the fugacity of positive (negative) charges.

Insert unity under the integral in the form

$$1 = \int D\rho(\vec{x}) \delta \left[\rho(\vec{x}) - \sum_{i=1}^{n+m} q_i \delta(\vec{x} - \vec{x}_i) \right],$$

to introduce the charge density $\rho(\vec{x})$. Writing the δ-function in terms of an additional integral over an auxiliary variable $\Phi(\vec{x})$, this can be recast into

$$1 = \int D\rho(\vec{x}) D\Phi(\vec{x}) e^{-i \int d\vec{x} \Phi(\vec{x}) [\rho(\vec{x}) - \sum_{i=1}^{n+m} q_i \delta(\vec{x} - \vec{x}_i)]}.$$

The expression in the exponent in Z may be written in terms of the charge density $\rho(\vec{x})$ as

$$\frac{1}{2T} \int d\vec{x} \, d\vec{y} \, \rho(\vec{x}) V(\vec{x} - \vec{y}) \rho(\vec{y}).$$

Since the exponent is quadratic in charge density, the functional integral over $\rho(\vec{x})$ is Gaussian and can be performed. After integration over particle trajectories, one finds

$$Z = \int D\Phi(\vec{x}) e^{-\frac{T}{2} \int d\vec{x} d\vec{y} \Phi(\vec{x}) V^{-1}(\vec{x} - \vec{y}) \Phi(\vec{y}) + \int d\vec{x} [(y_+ + y_-) \cos \Phi(\vec{x}) + i(y_+ - y_-) \sin \Phi(\vec{x})]}.$$

Since for the Coulomb potential $V^{-1}(\vec{x}) = -\delta(\vec{x}) \nabla^2 / (4\pi^2)$, identifying $\Phi(\vec{x})/2\pi = \theta(\vec{x})$ leads to $Z = Z_{sG}$, in the neutral case $y_+ = y_-$.

Problem 6.4 Using the Bogoliubov inequality $F \leq F_{var} = F_0 + \langle S - S_0 \rangle_0$, with $F = -\ln Z$ and S as the action, find the best variational quadratic approximation S_0 to the sine-Gordon action S_{sG} in Eq. (6.47). Show that at low fugacity the variational free energy has a continuous transition in two dimensions, and no transition in one and three dimensions. What phases should be expected in the one-dimensional and the three-dimensional Coulomb plasmas?

Solution Choosing

$$S_0 = \frac{1}{2} \int \frac{d\vec{k}}{(2\pi)^d} G_0^{-1}(k) \theta(\vec{k}) \theta(-\vec{k}),$$

with $G_0^{-1}(k) = Tk^2 + \Sigma(k)$, gives the variational free energy

$$F_{\text{var}} = \frac{1}{2} \int \frac{d\vec{k}}{(2\pi)^d} \left[-(\ln G_0(k)) + Tk^2 G_0(k) \right] - 2ye^{-\frac{4\pi^2}{2} \int \frac{d\vec{k}}{(2\pi)^d} G_0(k)},$$

where we set $J = a = 1$. Minimizing with respect to $G_0(k)$ leads to the equation

$$\Sigma(k) = 8\pi^2 y e^{-\frac{4\pi^2}{2} \int \frac{d\vec{k}}{(2\pi)^d} \frac{1}{Tk^2 + \Sigma(k)}},$$

which has a constant solution $\Sigma(k) = \Sigma$.

In $d = 2$ the equation for Σ becomes

$$\frac{\Sigma}{8\pi^2 y} = \left(\frac{\Sigma}{T\Lambda^2 + \Sigma} \right)^{\frac{\pi}{2T}}.$$

The right hand side of the equation approaches unity for large Σ, and vanishes for $\Sigma = 0$. For small fugacity the left hand side is a straight line with a large slope. Since for $T < \pi/2$ the right hand side is concave for small Σ, the only solution is $\Sigma = 0$, corresponding to the dielectric phase of bound dipoles. For $T > \pi/2$, on the other hand, the function on the right hand side becomes convex everywhere, and there are two solutions: $\Sigma = 0$ and $\Sigma \neq 0$. The latter one corresponds to the metallic phase with free charges. It is easy to check that the finite solution has a lower energy. Furthermore, as $T \to \pi/2$ from above the finite solution continuously approaches zero. In $d = 2$, F_{var} exhibits therefore a continuous phase transition at $T_c = \pi/2$.

In $d = 3$, the equation for Σ becomes

$$\frac{\Sigma}{8\pi^2 y} = e^{-\int_0^\Lambda \frac{k^2 dk}{Tk^2 + \Sigma}}.$$

Since the integral in the exponent remains finite when Σ approaches zero, in $d = 3$ there is a finite solution at arbitrary low fugacity and temperature. The Coulomb plasma is therefore expected to always be in the metallic phase in $d = 3$.

Finally, in $d = 1$, assuming a large cutoff Λ leads to

$$\frac{\Sigma}{8\pi^2 y} = e^{-\frac{\pi^2}{\sqrt{T\Sigma}}}.$$

The right hand side is always concave, and consequently the only solution is $\Sigma = 0$. The Coulomb plasma should thus always be in the dielectric phase in $d = 1$ at low fugacity.

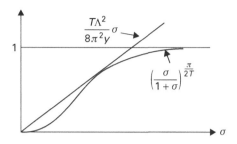

Figure 6.4 Graphical solution of the self-consistent equation at the point
of first-order transition between the dielectric and the metallic phases at
a large fugacity.

Problem 6.5 Using the variational free energy from the previous problem,
argue that the transition in a two-dimensional Coulomb plasma may become
discontinuous at large fugacity, and estimate the critical value of fugacity at
which this change occurs.

Solution Introducing the dimensionless $\sigma = \Sigma/(T\Lambda^2)$, the equation for Σ
in $d = 2$ becomes

$$\sigma \frac{T\Lambda^2}{8\pi^2 y} = \left(\frac{\sigma}{1+\sigma}\right)^{\frac{\pi}{2T}}.$$

Let us assume $T < \pi/2$. As the fugacity is increased at a fixed temperature,
there will appear a non-trivial solution $\sigma \neq 0$ when the straight line on the
left hand side first touches the curve described by the function on the right
(Fig. 6.4). So,

$$\frac{T\Lambda^2}{8\pi^2 y} = \frac{\pi}{2T} \left(\frac{\sigma}{1+\sigma}\right)^{\frac{\pi}{2T}-1} \frac{1}{(1+\sigma)^2}.$$

In combination with the equation for σ, we find

$$\sigma = \frac{\pi}{2T} - 1.$$

The critical value of fugacity at which the first-order transition occurs is given
by

$$\frac{y}{\Lambda^2} = \frac{1}{16\pi} \left(1 - \frac{2T}{\pi}\right)^{1-\frac{\pi}{2T}},$$

as depicted in Fig. 6.5.

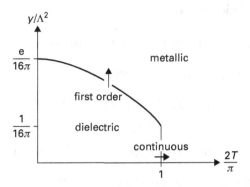

Figure 6.5 The phase diagram of the two-dimensional Coulomb plasma in the variational approximation. The thick line marks the discontinuous transition.

6.4 Renormalization of the sine-Gordon model

Once the XY model has been recast as the continuum sine-Gordon theory in Eq. (6.48) its critical behavior near the Kosterlitz–Thouless transition may be understood by using the standard momentum-shell transformation. In order to perform the integration over the fast modes we divide the dual field $\theta(\vec{x})$ as

$$\theta(\vec{x}) = \theta_<(\vec{x}) + \theta_>(\vec{x}), \tag{6.49}$$

where as usual $\theta_<(\vec{x})$ contains only the Fourier components with $k < \Lambda/b$, and $\theta_>(\vec{x})$ only with $\Lambda/b < k < \Lambda$. Since the interaction term in S_{sG} contains all powers of the dual field it will prove advantageous to work in real space. We will assume that the temperature is low enough so that $y \ll 1$ and use the perturbation theory in fugacity y. The partition function may then be written as

$$Z = Z_{0>} \int D\theta_< e^{-\frac{T}{2J} \int d\vec{x}(\nabla\theta_<(\vec{x}))^2} \left[1 + 2y \int d\vec{x} \, \cos(2\pi\theta_<(\vec{x}))\langle\cos(2\pi\theta_>(\vec{x}))\rangle_{0>} \right.$$

$$+ 2y^2 \int d\vec{x} \, d\vec{y} [\cos(2\pi\theta_<(\vec{x})) \cos(2\pi\theta_<(\vec{y}))\langle\cos(2\pi\theta_>(\vec{x})) \cos(2\pi\theta_>(\vec{y}))\rangle_{0>}$$

$$\left. + \sin(2\pi\theta_<(\vec{x})) \sin(2\pi\theta_<(\vec{y}))\langle\sin(2\pi\theta_>(\vec{x})) \sin(2\pi\theta_>(\vec{y}))\rangle_{0>}] + O(y^3) \right],$$

$$\tag{6.50}$$

where we have set $a = 1$, and used the fact that an average of an odd number of fields over the Gaussian action vanishes, so that

$$\langle\sin(2\pi\theta_>(\vec{x}))\rangle_{0>} = \langle\sin(2\pi\theta_>(\vec{x})) \cos(2\pi\theta_>(\vec{y}))\rangle_{0>} = 0. \tag{6.51}$$

Defining the correlation function

$$g_>(\vec{x}) = (2\pi)^2 \langle \theta_>(\vec{x})\theta_>(0)\rangle_{0>}, \tag{6.52}$$

the averages appearing in Eq. (6.50) become

$$\langle \cos(2\pi\theta_>(\vec{x}))\rangle_{0>} = e^{-\frac{1}{2}g_>(0)}, \tag{6.53}$$

$$\langle \cos(2\pi\theta_>(\vec{x}))\cos(2\pi\theta_>(\vec{y}))\rangle_{0>} = e^{-g_>(0)}\cosh(g_>(\vec{x}-\vec{y})), \tag{6.54}$$

$$\langle \sin(2\pi\theta_>(\vec{x}))\sin(2\pi\theta_>(\vec{y}))\rangle_{0>} = e^{-g_>(0)}\sinh(g_>(\vec{x}-\vec{y})). \tag{6.55}$$

When re-exponentiated, the terms in Eq. 6.50 determine the action for the slow modes:

$$S_< = \frac{T}{2J}\int d\vec{x}(\nabla\theta_<(\vec{x}))^2 - 2ye^{-\frac{1}{2}g_>(0)}\int d\vec{x}\cos(2\pi\theta(\vec{x}))$$

$$- y^2 e^{-g_>(0)}\int d\vec{x}\, d\vec{y}\Big[\cos(2\pi(\theta_<(\vec{x})+\theta_<(\vec{y})))\left(e^{-g_>(\vec{x}-\vec{y})}-1\right)$$

$$+ \cos(2\pi(\theta_<(\vec{x})-\theta_<(\vec{y})))\left(e^{g_>(\vec{x}-\vec{y})}-1\right)\Big] + O(y^3). \tag{6.56}$$

To the first order in fugacity, the result of the momentum-shell integration is therefore only the change of fugacity,

$$y \to y(b) = b^2 y e^{-\frac{1}{2}g_>(0)} + O(y^2), \tag{6.57}$$

where

$$g_>(0) = \frac{2\pi}{T}\int_{\Lambda/b}^{\Lambda}\frac{dk}{k} = \frac{2\pi}{T}\ln(b). \tag{6.58}$$

We have set $J = 1$ here as well for convenience. The factor b^2 in $y(b)$ compensates for the change in the short-distance cutoff $a \to a(b) = ba$, which is the real-space equivalent of the usual rescaling of the momentum cutoff.

The renormalization of temperature derives from the second-order terms in $S_<$. Expanding $\theta_<(\vec{y})$ around $\vec{x}$, to the leading order in gradients one can write

$$\cos(2\pi(\theta_<(\vec{x}) - \theta_<(\vec{y}))) = 1 - \frac{1}{2}(2\pi(\vec{y}-\vec{x})\cdot\nabla\theta_<(\vec{x}))^2 + O(\theta_<\nabla^3\theta_<), \tag{6.59}$$

and

$$\cos(2\pi(\theta_<(\vec{x}) + \theta_<(\vec{y})))$$
$$= \cos(4\pi\theta_<(\vec{x})) - 2\pi\sin(4\pi\theta_<(\vec{x}))[(\vec{y}-\vec{x})\cdot\nabla\theta_<(\vec{x})] + \cdots. \tag{6.60}$$

The second term in Eq. (6.59) therefore renormalizes the temperature. When inserted back into Eq. (6.56) it yields

$$T \to T(b) = T + 2\pi^2 y^2 e^{-g_>(0)} \int d\vec{x} \left(e^{g_>(\vec{x})} - 1\right)\vec{x}^2. \tag{6.61}$$

Equation (6.59) also generates the higher-order gradients of the dual field, which may be dropped as irrelevant at low fugacity. Equation (6.60), on the other hand, introduces the next "harmonic", $\cos(4\pi\theta_<)$, into the action. We will see that such higher harmonics may also be omitted as irrelevant close to the transition. The fugacity is thus not renormalized by the second-order terms, and Eq. (6.57) is in fact correct to $O(y^3)$.

We may attempt to evaluate $g_>(x)$ in Eq. (6.61) similarly to $g_>(0)$:

$$g_>(\vec{x}) = \frac{1}{T} \int_{\Lambda/b}^{\Lambda} \frac{dk}{k} \int_0^{2\pi} d\alpha e^{ikx\cos(\alpha)} = \frac{2\pi}{T} \ln(b) J_0(\Lambda x) + O((\ln(b))^2). \tag{6.62}$$

The integral in $T(b)$ to the first order in $\ln(b)$ is therefore

$$\sim \ln(b) \int d\vec{x}\, x^2 J_0(\Lambda x), \tag{6.63}$$

which, since $J_0(z) \approx \cos(z - \pi/4)/\sqrt{z}$ for large z, is divergent in the thermodynamic limit. At first this must seem like a disaster, particularly in view of the effort involved to come to this point. Fortunately, this divergence is only an artifact of the way we implemented the cutoff in the calculation. The integral in question may be understood as giving the change in polarizability of the system of bound dipoles when the minimal size of a dipole is increased from $a \sim 1/\Lambda$ into $a \sim b/\Lambda$. Since for small fugacity and at low temperature such polarizability is certainly finite, its change should be finite as well.

This failure of our straightforward computation serves as a warning that there is nothing automatic about the renormalization group, and more often than not new problems require non-trivial reformulations guided by physical understanding. In the present case the difficulty may be eliminated by the introduction of the "soft" cutoff as follows. First, rewrite

$$g_>(\vec{x}) = \int_0^{\Lambda} \frac{e^{i\vec{q}\cdot\vec{x}}}{Tq^2} d\vec{q} - \int_0^{\Lambda/b} \frac{e^{i\vec{q}\cdot\vec{x}}}{Tq^2} d\vec{q}, \tag{6.64}$$

and then modify

$$\int_0^{\Lambda} d\vec{q} \to \int_0^{\infty} d\vec{q} \frac{\Lambda^2}{q^2 + \Lambda^2}. \tag{6.65}$$

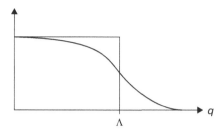

Figure 6.6 Sharp and smooth cutoffs in the momentum integrals.

This replaces the step-like cutoff in the momentum space with a smooth function, as depicted in Fig. 6.6.

With this replacement and after some rearrangement of the terms, one finds

$$g_>(x) = \frac{2\pi}{T}\left(\int_0^\infty dq \frac{q J_0(qx)}{q^2 + (\Lambda/b)^2} - \int_0^\infty dq \frac{q J_0(qx)}{q^2 + \Lambda^2}\right)$$

$$= \frac{2\pi}{T}(K_0(x\Lambda/b) - K_0(x\Lambda))$$

$$= -\frac{2\pi x\Lambda \ln(b)}{T}\frac{dK_0(z)}{dz}\Big|_{z=x\Lambda} + O((\ln(b))^2), \tag{6.66}$$

where $K_0(z)$ is the modified Bessel function. Since $K_0(z) = -\ln(z) + \text{const}$ for $z \ll 1$, for $g_>(0)$ we reproduce the result in Eq. (6.58). The renormalized temperature, however, now becomes

$$T(b) = T - 2\pi^2 y^2 \int d\vec{x}\, x^2 \frac{2\pi x\Lambda}{T}\frac{dK_0(z)}{dz}\Big|_{z=x\Lambda} \ln(b) + O(y^2 \ln(b)^2), \tag{6.67}$$

and perfectly finite, since $K_0(z) \approx (\pi/2z)^{1/2}e^{-z}$ for $z \gg 1$. Performing the integral in the last equation over the whole plane, one obtains

$$T(b) = T + \frac{1}{2T}\left(\frac{y(4\pi)^2}{\Lambda^2}\right)^2 \ln(b). \tag{6.68}$$

Defining the dimensionless fugacity as $\hat{y} = (4\pi)^2 y/\Lambda^2$, the flow of the two couplings in the sine-Gordon theory may be put into the differential form:

$$\frac{d\hat{y}}{d\ln(b)} = \left(2 - \frac{\pi}{T}\right)\hat{y} + O(\hat{y}^3), \tag{6.69}$$

$$\frac{dT}{d\ln(b)} = \frac{\hat{y}^2}{2T} + O(\hat{y}^4). \tag{6.70}$$

The flow diagram is depicted in Fig. 6.7. Whereas the temperature always increases under renormalization, a small fugacity is irrelevant for

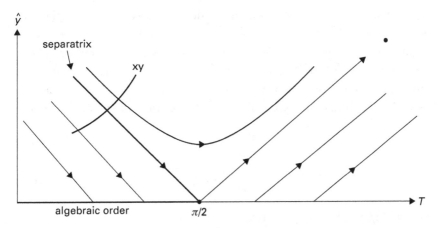

Figure 6.7 Renormalization group flow of the temperature and the fugacity in the sine-Gordon model. The thick line separates the flows towards the line of fixed points at $T < T_{KT}$ and $\hat{y} = 0$ that represent the algebraically ordered superfluid phase from the flow towards a high-temperature high-fugacity sink, representing the exponentially disordered phase. The XY line represents the set of initial values at different temperatures in the XY model.

$T < T_{KT} = \pi/2$, and relevant otherwise. At a finite fugacity the critical temperature is $T_c < T_{KT}$, and is determined by the separatrix in the flow diagram. The result is therefore qualitatively the same as in the mean-field theory. This is because the fugacity renormalizes to zero for $T < T_c$, and does so even right at the separatrix, when $T = T_c$. The Kosterlitz–Thouless critical point is thus non-interacting in the dual (sine-Gordon) formulation, so that the transition is indeed due to unbinding of the single largest vortex-antivortex pair, just as the simplest energy–entropy argument would suggest.

Since vortices are still bound at all $T \leq T_c$, we may calculate the correlation function in the ordered phase of the XY model exactly at large distances by using the Gaussian approximation, only with T replaced by $T_\infty = T(b \to \infty)$. Right at $T = T_c$ the temperature flows to T_{KT} and

$$\left\langle e^{i\phi(\vec{x})} e^{-i\phi(0)} \right\rangle \approx |\vec{x}|^{-\frac{T_{KT}}{2\pi}}. \tag{6.71}$$

Since the above power-law defines the anomalous dimension η by $(T_{KT}/2\pi) = d - 2 + \eta$, we find that at the Kosterlitz–Thouless transition

$$\eta = \frac{1}{4}. \tag{6.72}$$

The correlation length, on the other hand, does not diverge as a simple power law, since the stability matrix vanishes at $y = 0$ and $T = T_{KT}$. We can still determine the divergence of correlation length, however, by finding how fast the neighborhood of the critical point is escaped under renormalization. Let us introduce more convenient variables for this purpose as $z = y^2$ and $x = 2 - \pi/T$, so that near the Kosterlitz–Thouless critical point at $x = z = 0$,

$$\frac{dz}{d\ln(b)} = 2xz, \tag{6.73}$$

$$\frac{dx}{d\ln(b)} = cz, \tag{6.74}$$

with $c = (2/\pi)^2$. Dividing the two,

$$\frac{dz}{dx} = \frac{2}{c}x, \tag{6.75}$$

so that $z = (x^2/c) + \sigma$. Let us assume the initial condition that at $b = 1$, $x = 0$ and $z = \sigma$, and determine the value of the rescaling parameter b at which $z(b) = 1$. The parameter σ may then be understood as the deviation from the critical point: $\sigma \sim (T - T_c)/T_c$. Integrating Eq. (6.73),

$$\ln b = \frac{1}{2\sqrt{c}} \int_\sigma^1 \frac{dz}{z\sqrt{z-\sigma}} = \frac{\pi^2}{4\sqrt{\sigma}}, \tag{6.76}$$

for $\sigma \ll 1$. The correlation length $\xi \sim b$, and therefore

$$\xi \sim e^{C\sqrt{\frac{T_c}{T-T_c}}}, \tag{6.77}$$

where the constant C is a non-universal number. At T_c the correlation length therefore exhibits an *essential singularity*. From the flow diagram and our definition of the correlation length it immediately follows that for $T < T_c$, $\xi = \infty$.

Since the Kosterlitz–Thouless critical point is at $y = 0$ one may worry that the fugacity may be a dangerously irrelevant coupling. This is not so: if $y < 0$,

$$-|y|\cos(2\pi\theta) = |y|\cos(2\pi\theta + \pi), \tag{6.78}$$

so by changing the dual variable globally as $\theta(\vec{x}) + (1/2) \rightarrow \theta(\vec{x})$ we see that the sine-Gordon partition function is an even function of fugacity. The model is therefore well-defined for $y < 0$, and in particular hyperscaling is satisfied. The free energy per unit area scales therefore like $f \sim \xi^{-2}$. The free energy, and consequently the specific heat, thus also have an essential singularity at $T = T_c$. The usual critical exponents ν, α, β, and γ cannot therefore be defined for the Kosterlitz–Thouless transition. We may nevertheless define the exponent δ, which from Eq. (3.55) is then

$$\delta = 15. \tag{6.79}$$

The invariance of the sine-Gordon theory under the change of sign of fugacity also implies that the right hand sides in Eqs. (6.69) and (6.70) contain only odd and even powers of $\hat{y}$, respectively.

Since the Kosterlitz–Thouless critical point is non-interacting in the dual formulation, the relevance of any additional couplings in S_{sG} is easily determined. Let us, for example, add to the sine-Gordon action the term

$$S' = 2 \sum_{n=2}^{\infty} y_n \int d\vec{x}\, \cos(2\pi n\theta(\vec{x})). \tag{6.80}$$

The nth term in the sum then represents vortices with vorticity n. The energy–entropy argument suggests that vortices with higher vorticity should still be bound at the Kosterlitz–Thouless transition, and indeed the renormalization group analysis confirms this expectation. To the lowest order it follows that

$$\frac{dy_n}{d\ln(b)} = \left(2 - \frac{n^2\pi}{T}\right) y_n, \tag{6.81}$$

so that at $T = T_{KT}$ all the harmonics with $n > 1$ are irrelevant. This justifies the omission of the term in Eq. (6.60).

To summarize, after a series of approximations we found that the two-dimensional XY model may be mapped onto the sine-Gordon theory with a non-interacting critical point. This critical point controls the Kosterlitz–Thouless phase transition in the equivalent neutral two-dimensional Coulomb plasma of unit charges, representing vortices and antivortices in the XY model. Our approach was based on the perturbation theory in vortex fugacity, which may always be assumed to be low at low temperatures. The transition, however, was found to be at $T \approx T_{KT} = 2J/\pi$, so the relevance of our results for the specific XY model still hinges on the smallness of the vortex fugacity not only at low temperatures, but even near the Kosterlitz–Thouless transition temperature. To check the consistency of our approach one thus needs to compute the appropriate value of the chemical potential, or the core energy, of the vortex in the XY model. Let us define it as the finite part of the total energy of the vortex with vorticity q:

$$H_v = \pi J q^2 \ln \frac{R}{a} - \mu q^2, \tag{6.82}$$

where a is exactly the lattice spacing and R the size of the system. This quantity clearly depends on the details of the model on the scale of few lattice spacings. It essentially measures the cost in energy near the center of the vortex, where the continuum approximation, appropriate far from the center, fails. An excellent approximation[1] for the energy of a unit vortex in the XY model for all $R \geq a$ is

$$H_v = \pi J \ln \frac{R}{\tau}, \tag{6.83}$$

with $a/\tau = 2\sqrt{2}e^{0.577}$. This yields $\mu/J \approx -\pi^2/2$. Near the Kosterlitz–Thouless transition temperature we thus find $y \approx e^{-\pi} \approx 0.04$, just about small enough for the whole approach to be self-consistent. Tuning the temperature in the XY model changes therefore both the parameters T and y in the sine-Gordon theory along the "XY line" of initial values in the flow diagram in Fig. 6.7. At the point of intersection with the separatrix the XY model goes through the Kosterlitz–Thouless phase transition governed by the zero-fugacity critical point of the sine-Gordon theory.

6.5 Universal jump of superfluid density

The fixed point values of the couplings are typically non-universal and without direct physical significance. We are therefore usually interested in critical exponents or amplitude ratios, which are universal and measurable. An interesting feature of Eqs. (6.69) and (6.70) is that the fixed point value of one of the couplings has in fact direct physical significance. Let us restore J and define the coupling

$$K = \frac{J}{T}. \tag{6.84}$$

The result of the renormalization group analysis is that

$$\lim_{\Delta T \to 0+} \lim_{b \to \infty} [K(b, T_c - \Delta T) - K(b, T_c + \Delta T)] = \frac{2}{\pi}. \tag{6.85}$$

The asymptotic value of the coupling K drops to zero discontinuously at $T = T_c$. On the other hand, since vortices are irrelevant for $T \leq T_c$ we may treat the XY model in the Gaussian approximation. Comparing the coefficients in front of the $(\nabla \phi)^2$ term and restoring back all the constants, from Eq. (2.36)

[1] J. M. Kosterlitz, *Journal of Physics C* **7**, 1046 (1974)

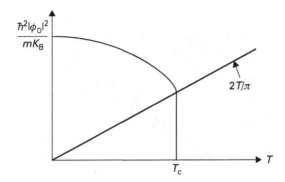

Figure 6.8 Temperature dependence of the superfluid density in two dimensions. The straight line is the universal relation in Eq. (6.87).

we find that

$$K(b = 1, T) = \frac{\hbar^2 |\Phi_0(T)|^2}{mk_B T}, \qquad (6.86)$$

where $|\Phi_0(T)|^2$ is the bare superfluid areal density, and m is the mass of particles. The flow of the coupling K may therefore be understood as the renormalization of the superfluid density from its bare value at $b = 1$ to its physical value at $b = \infty$. In particular, Eq. (6.85) implies that

$$\frac{\hbar^2 |\Phi_0(T_c^-)|^2}{mk_B T_c} = \frac{2}{\pi}. \qquad (6.87)$$

Although neither the discontinuity in superfluid density nor the critical temperature are universal by themselves, their ratio is universal at the Kosterlitz–Thouless transition. The superfluid density in a two-dimensional superfluid system should therefore depend on temperature as depicted schematically in Fig. 6.8.

The universal jump in superfluid density has indeed been observed. After a thin film of ^{4}He is absorbed on a substrate, the film is wound as a spiral on the axis of a torsional oscillator. When ^{4}He becomes superfluid the moment of inertia of the oscillator drops, which causes a decrease in the period of oscillations. This decrease, on the other hand, determines the superfluid density. The results of experiment are in very good agreement with the prediction of the Kosterlitz–Thouless theory (see Fig. 6.9)

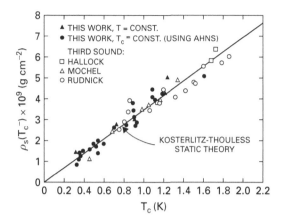

Figure 6.9 Measurements of the superfluid density at T_c vs. T_c in two dimensional ^{4}He. The straight line is Eq. (6.87). (Reprinted with permission from D. Bishop and J. Reppy, *Physical Review Letters* **40**, 1727 (1978). Copyright 1978 by the American Physical Society.)

6.6 Heisenberg model

It is interesting to contrast the XY with the Heisenberg model in two dimensions, and understand the role of topology of the order parameter in the Kosterlitz–Thouless transition. Although one can identify topologically nontrivial configurations in the Heisenberg model as well, we will see that there is no equivalent of the Kosterlitz–Thouless transition in this case.

Two features of the XY model are crucial for the existence of the Kosterlitz–Thouless transition: (1) vortices are local minima of the energy, and are therefore decoupled from the spin waves, and (2) the energy of a vortex is logarithmically divergent in the thermodynamic limit, so at low temperatures there can not be any free vortices in the system. The first condition is what makes it possible to consider the system of vortices separately, as done in the Coulomb gas representation of the XY model, for example. The reason behind it is basically topological, and we turn to some elementary topological considerations next.

For the two-dimensional XY model we may define the *winding number* of a configuration as

$$q = \frac{1}{2\pi} \oint d\vec{x} \cdot \nabla \phi(\vec{x}), \tag{6.88}$$

where we assumed a configuration $\phi(\vec{x})$ which is smooth everywhere except possibly at some points, together with the continuum limit. The contour of integration is the boundary of the system. Due to the periodicity of the phase the above integral is an integer, and it simply measures the net number of vortices in the system. Since the function $\phi(\vec{x})$ for $\vec{x}$ at the boundary of a two-dimensional region maps a circle onto another circle, the above integral counts how many times such a map wraps around the circle in ϕ-space as $\vec{x}$ goes around the system once. The integral q is then *topologically invariant*: if we define a class of continuous functions $\phi(\vec{x}, t)$, so that the initial point $\phi(\vec{x}_{in}, t)$ and the final point $\phi(\vec{x}_{fin}, t)$ are independent of t, the integral q is the same for all the functions in the same class. Such a class is called a *homotopy class*. This implies that there is no function $\phi(\vec{x}, t)$ continuous in t so that $\phi(\vec{x}, 0) = \text{const}$ is the ground state, and $\phi(\vec{x}, 1)$ is the vortex, since these two configurations belong to different homotopy classes, and thus have different winding numbers. We can thus imagine the space of configurations of the XY model being partitioned into different homotopy classes distinguished by the value of their winding numbers, with each class having its own local minimum in energy. For $q = 0$ this minimum is obviously the ground state, while for $q = 1$ it is the vortex in Eq. (6.4).

Local stability of a configuration follows therefore from it being topologically distinct from the ground state. Consider next the Heisenberg non-linear σ-model in two dimensions:

$$S_H = \frac{J}{2T} \int d\vec{x} (\nabla \vec{n}(\vec{x}))^2, \qquad (6.89)$$

with $\vec{n}(x)$ as a three-dimensional vector of unit length. In spherical coordinates we can parameterize $\vec{n}$ by two angles, $0 < \phi < 2\pi$ and $0 < \theta < \pi$. Taking $\vec{x}$ to again lie at the boundary of the system, the function $\vec{n}(\vec{x})$ now provides a mapping of a circle onto the surface of a unit sphere. We may again define the winding number in Eq. (6.88), but it is no longer topologically invariant; any closed path on a sphere can be deformed continuously into a point simply by allowing the vectors $\vec{n}(\vec{x})$ to vary in the third direction. Vortices are therefore no longer locally stable configurations, since tilting the vectors slightly off the plane may obviously lower the energy. This is in accord with the above topological argument which tells that a vortex can now be continuously deformed into the ground state, with the energy being decreased monotonically in the process. In other words, in the configurational space of the Heisenberg model a vortex is simply a higher energy state in the same homotopy class with the ground state.

There exists, however, a different topologically distinct configuration in the Heisenberg model. Consider the following integral:

$$N = \frac{1}{4\pi} \int \int dx\, dy [\hat{n}(\vec{x}) \cdot (\partial_x \hat{n}(\vec{x}) \times \partial_y \hat{n}(\vec{x}))], \qquad (6.90)$$

which measures how many times the mapping $\vec{n}(\vec{x})$ of a plane onto a unit sphere encloses the sphere. Let us define the *skyrmion* configuration with $N = 1$: $\phi = \arctan(y/x)$, and θ a continuous function of $r = (x^2 + y^2)^{1/2}$ which vanishes at the origin and equals π for $r > a$, with a being some length. The action in Eq. (6.89) for a slowly varying configuration in terms of the angles is

$$S_H = \frac{J}{2T} \int \int dx\, dy [(\nabla \theta)^2 + \sin^2 \theta (\nabla \phi)^2], \qquad (6.91)$$

which for the skyrmion becomes

$$S_{sk} = \frac{\pi J}{T} \int_0^a \left[\left(\frac{d\theta}{dr} \right)^2 + \frac{1}{r^2} \sin^2 \theta \right] r\, dr. \qquad (6.92)$$

For the XY model the angle $\theta = \pi/2$, and the second term in Eq. (6.91) led to the divergent energy of the vortex both in the upper and the lower limit of the integration. The former denoted the divergence in the thermodynamic limit whereas the latter signaled the breakdown of the continuum approximation for the energy near the center of the vortex. For the skyrmion, in contrast, the action is finite in both limits. If we choose $\theta \approx \pi r/a$ for $0 < r < a$, for example, the action becomes

$$S_{sk} = \frac{\pi J}{T} \left[\frac{\pi^2}{2} + \int_0^\pi \frac{\sin^2 t}{t} dt \right], \qquad (6.93)$$

and independent of the length a. Taking a to be small, in fact, would make the skyrmion different from the ground state with $N = 0$ only in the small disc of radius a. Of course, this would not everywhere be a slowly varying configuration and the continuum approximation for the energy in Eq. (6.91) would no longer apply. Nevertheless, the energy of such a configuration will obviously be finite. Skyrmions are therefore expected to be free at any finite temperature, in contrast to the energetically costly vortices in the XY model.

To summarize, the XY model in two dimensions is special in that the topologically distinct configurations have diverging energies in the thermodynamic limit. Although the two-dimensional Heisenberg model has skyrmions as its own topologically distinct configurations, in contrast to vortices they have finite energies. The Heisenberg model is therefore always disordered in two dimensions.

Problem 6.6 Show that the superfluid density has a square-root cusp at $T = T_c$ in two dimensions.

Solution From Eq. (6.75) we found that

$$y^2 = \frac{\pi^4}{4}\left(\frac{2}{\pi} - K\right)^2 + \sigma$$

describes the renormalization trajectories, with $K = J/T$. When $\sigma < 0$ the system is in the ordered phase, with

$$K = \frac{2}{\pi}\left(1 + \frac{1}{\pi}\sqrt{-\sigma}\right)$$

and $y = 0$ in the limit $b \to \infty$. Since $-\sigma \propto T_c - T$, the superfluid density exhibits a square-root cusp at the critical temperature.

Problem 6.7 The quantum superfluid–insulator transition in a random potential at $T = 0$ and in one dimension may be described by the action

$$S_{SI}[\theta] = K \int dx\, d\tau \sum_{i=1}^{N}\left[(\partial_\tau\theta_i(x, \tau))^2 + c^2(\partial_x\theta_i(x, \tau))^2\right]$$

$$- W \sum_{i,j=1}^{N} \int dx\, d\tau\, d\tau'\, \cos 2[\theta_i(x, \tau) - \theta_j(x, \tau')],$$

in the limit $N \to 0$. The coupling W measures the strength of the impurity potential, $c = \sqrt{\rho_s/(m\kappa)}$ is the velocity of second sound in the superfluid, and $K = m/(2\pi^2\hbar\rho_s)$, where ρ_s is the superfluid density and κ the compressibility. Find the behavior of the velocity c at the transition.

Solution Let us separate the slow and the fast modes as $\theta_i(x, \tau) = \theta_{i<}(x, \tau) + \theta_{i>}(x, \tau)$, where

$$\theta_{i<}(x, \tau) = \int_{-\Lambda/b}^{\Lambda/b}\frac{dk}{2\pi}\int_{-\infty}^{\infty}\frac{d\omega}{2\pi}\theta_i(k, \omega)e^{ikx+i\omega\tau}.$$

To the first order in W the integration over $\theta_{i>}$ leads to the action $S_< = S_{SI}[\theta_<]$ for the slow modes, except for the additional factor

$$\langle\cos 2(\theta_{i>}(x, \tau) - \theta_{j>}(x, \tau'))\rangle_{0,>} = \exp\left(-\frac{\Lambda\ln b}{\pi^2 K}\int_{-\infty}^{\infty}d\omega\frac{1 - \delta_{ij}e^{i\omega(\tau-\tau')}}{c^2\Lambda^2 + \omega^2}\right)$$

under the integral in the last term, in analogy to the Eq. (6.50). Expanding the last term in $S_<$ in powers of frequency and of fields one finds

$$\frac{d\hat{W}}{d\ln b} = \left(3 - \frac{1}{\pi K c}\right)\hat{W},$$

$$\frac{dK}{d\ln b} = \frac{8\hat{W}}{\pi K c^4},$$

$$\frac{dK c^2}{d\ln b} = 0,$$

where $\hat{W} = W/\Lambda^3$. Due to non-locality of the last term in S_{SI} in imaginary time the coefficient of $(\partial_\tau \theta_i)^2$ becomes renormalized already to the lowest order in W.

Since the combination $K c^2 = 1/(2\pi^2 \hbar \kappa)$ remains constant under renormalization, the velocity c decreases as the transition is approached from the superfluid side, and vanishes in the insulator. At the transition $1/(Kc) = 3\pi$, so the velocity of second sound has a universal (relative to compressibility) discontinuity

$$\hbar \kappa \Delta c = \frac{3}{2\pi}$$

at the superconductor–insulator transition in one dimension.

Problem 6.8* Consider the two-dimensional XY model in Eq. (6.29) in presence of the symmetry breaking perturbation

$$S_{\mathrm{XY}} = -\frac{1}{T}\sum_{i,\hat{v}} \cos(\phi_i - \phi_{i+\hat{\mu}}) - h\sum_i \cos(p\phi_i),$$

with integer p. The last term reduces the continuous $U(1)$ symmetry of the XY model to a discrete Z_p symmetry under $\phi_i \to \phi_i + (2\pi/p)$. Determine the values of p for which there still exists a Kosterlitz–Thouless transition at small h.

Solution Assume $T < T_c$, where T_c is the Kosterlitz–Thouless transition temperature at $h = 0$. The first term in S_{XY} may then be replaced by an effective Gaussian term, and

$$S_{\mathrm{XY}} \to \frac{1}{2T_\infty} \int d^2\vec{x}(\nabla\phi)^2 - h \int d^2\vec{x}\,\cos(p\phi)$$

in the continuum notation, with the small ultraviolet cutoff Λ/b, and with $T_\infty = T(b \to \infty) > T$ as the renormalized temperature in the ordered phase for $h = 0$. Comparison with the sine-Gordon theory in Eq. (6.48) immediately implies the flow of the symmetry breaking field h:

$$\frac{dh}{d \ln(b)} = h \left(2 - \left(\frac{p}{2\pi} \right)^2 \pi T_\infty \right) + O(h^2),$$

in analogy with Eq. (6.81). Symmetry breaking perturbation is therefore irrelevant for $T_\infty > 8\pi/p^2$. Since $T_\infty < T_{KT} = \pi/2$, for $p > 4$ there exists an interval of temperatures below T_c for which h is still irrelevant. For $p > 4$, as temperature is decreased there is first a Kosterlitz–Thouless transition at T_c from the disordered into the algebraically ordered phase, and then another transition at a lower temperature into a long-range ordered phase that breaks the Z_p symmetry.

Problem 6.9* Find the value of the anomalous dimension η at the transition from algebraically ordered into long-range ordered phase in the previous problem, for $p = 6$.

Solution For $p > 4$, h becomes relevant in the algebraically ordered phase at $T_\infty = 8\pi/p^2$. Since the anomalous dimension is $\eta = T_\infty/(2\pi)$, for $p = 6$ we find $\eta = 1/9$.

Problem 6.10* Find the minimal extension of the Kosterlitz–Thouless flow equations to $d = 2 + \epsilon$ dimensions, and use it to obtain the correlation length exponent for the XY model near $d = 2$.

Solution In $d = 2 + \epsilon$, the combination J/T scales like $\sim b^\epsilon$. Setting $J = 1$ the Kosterlitz recursion relations become

$$\frac{d\hat{y}}{d \ln(b)} = \left(2 - \frac{\pi}{T} \right) \hat{y},$$

$$\frac{dT}{d \ln(b)} = -\epsilon T + \frac{\hat{y}^2}{2T}.$$

Instead of the critical line as in Fig. 6.7 there is now a standard critical point at $T_c = \pi/2$ and $y_c = \pi \sqrt{\epsilon/2}$. The correlation length exponent then reads

$$\nu = \frac{1}{2\sqrt{\epsilon}} + O(1).$$

7

Duality in higher dimensions

Duality transformations are explored in dimensions higher than two. The three-dimensional XY model is mapped onto the frozen lattice superconductor, and various duality relations between the correlation functions in two theories are established. The transition in the XY model is shown to correspond to confinement of the magnetic flux in the lattice superconductor. In four dimensions the lattice superconductor is shown to be dual to the pure compact electrodynamics.

7.1 Frozen lattice superconductor

We now turn our attention to the three-dimensional XY model defined by the same Eq. (6.29), only with the phases residing on sites of a three-dimensional quadratic lattice, so that $\hat{\mu} = \hat{x}, \hat{y}, \hat{z}$. Repeating the exact same steps leads to the current representation of the XY model as in Eq. (6.36), with $\vec{n}_i$ now being a divergence-free three-dimensional integer vector.

In two dimensions, the constraint on $\{\vec{n}_i\}$ in Eq. (6.36) was resolved by the introduction of a unique (up to an overall constant) set of variables $\{m_i\}$ on the dual lattice. This was the lattice version of the continuum relation $\vec{n}(\vec{r}) = (\partial_y m(\vec{r}), -\partial_x m(\vec{r}))$ for continuous variables, which guarantees that $\nabla \cdot \vec{n}(\vec{r}) = 0$ for arbitrary $m(\vec{r})$ in two dimensions. Obviously, the shift $m(\vec{r}) \rightarrow m(\vec{r}) + c$ with c as a constant leaves $\vec{n}(\vec{r})$ invariant. Here lies an important difference in three dimensions. From the constraint $\nabla \cdot \vec{n}(\vec{r}) = 0$ it follows that $\vec{n}(\vec{r}) = \nabla \times \vec{m}(\vec{r})$, where $\vec{m}(\vec{r})$ is an arbitrary vector. Furthermore, instead of the constant shift in two dimensions, in three dimensions the transformation $\vec{m}(\vec{r}) \rightarrow \vec{m}(\vec{r}) + \nabla \chi(\vec{r})$ with arbitrary function $\chi(\vec{r})$ leaves $\vec{n}(\vec{r})$ invariant. This hidden *gauge invariance* of the three-dimensional XY model will lead to a rather different form of the dual theory in three dimensions.

147

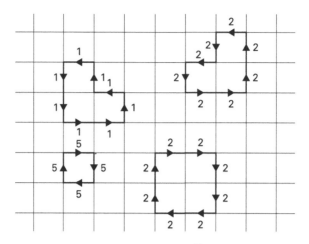

Figure 7.1 An example of vortex loops $\{\vec{M}_i\}$ on a quadratic lattice.

Returning to the discrete lattice, we may therefore write $\vec{n}_i = \Delta \times \vec{m}_i$, with the lattice version of the curl as

$$n_{i,\mu} = \sum_{\nu,\sigma} \epsilon_{\mu\nu\sigma}(m_{i-\hat{\sigma},\sigma} - m_{i-\hat{\nu}-\hat{\sigma},\sigma}). \tag{7.1}$$

Turning the integers $\{\vec{m}_i\}$ into real variables, the current representation of the Villain approximation of the three-dimensional XY model is written as

$$Z_V = \int_{-\infty}^{\infty} \prod_i d\vec{a}_i \sum_{\{\vec{p}_i\}} e^{-\frac{T}{2}\sum_i(\Delta\times\vec{a}_i)^2 + i2\pi \sum_i(\Delta\times\vec{a}_i)\cdot\vec{p}_i}, \tag{7.2}$$

where again the summation over integers $\{p_{i,\mu}\}$ turns the variables $\vec{a}_i$ back to integers, and we set $J = 1$. Finally, the last equation can be rewritten as

$$Z_V = \int_{-\infty}^{\infty} \prod_i d\vec{a}_i \sum_{\{\vec{M}_i\}} e^{-\frac{T}{2}\sum_i(\Delta\times\vec{a}_i)^2 - i2\pi \sum_i \vec{M}_i\cdot\vec{a}_i} \delta(\Delta \cdot \vec{M}_i), \tag{7.3}$$

with $\vec{M}_i = \Delta \times \vec{p}_i$.

Equation (7.3) is reminiscent of Eq. (6.43) in two dimensions except for the constraint on the integer vectors $\{\vec{M}_i\}$. The constraint implies that all configurations of $\{\vec{M}_i\}$ are superpositions of closed loops on the lattice, as depicted in Fig. 7.1, which will be called *vortex loops*. Similarly to two dimensions, the integration over the gauge field $\vec{a}_i$ would induce the Coulomb ($\sim 1/r$) interaction between the loop segments, just as the integration over the variables θ_i in Eq. (6.43) introduced logarithmic interaction between vortices in two dimensions.

The vortex-loop representation in Eq. (7.3) is an exact rewriting of the three-dimensional XY model in the Villain approximation. It is illuminating to transform Z_V even further by adding an infinitesimal chemical potential for the vortex loops in the exponent of Eq. (7.3):

$$-\frac{t}{2}\sum_i \vec{M}_i^2. \tag{7.4}$$

As discussed in Section 6.3, such a term should be irrelevant, since it only alters slightly the short-range part of the long-range interactions between vortex loop segments. With the chemical potential, however, the partition function in Eq. (7.3) may be rewritten as

$$Z_V = \lim_{t \to 0} \int_{-\infty}^{\infty} \prod_i [d\vec{a}_i \, d\theta_i \, d\vec{X}_i]$$
$$\times \sum_{\{\vec{n}_i\}} e^{-\frac{T}{2}\sum_i (\Delta \times \vec{a}_i)^2 - i2\pi \sum_i \vec{X}_i \cdot \vec{a}_i - \frac{t}{2}\sum_i \vec{X}_i^2 + i\sum_{i,\mu} X_{i,\mu}(\theta_{i+\hat{\mu}} - \theta_i - 2\pi n_{i,\mu})}. \tag{7.5}$$

The integration over $\{\theta_i\}$ makes $\{\vec{X}_i\}$ divergenceless, and the summation over new integers $\{\vec{n}_i\}$ restricts $\{\vec{X}_i\}$ to integer values, bringing back the form in Eq. (7.3). Integrating over $\{\vec{X}_i\}$, on the other hand, yields

$$Z_V = \lim_{t \to 0} \int_{-\infty}^{\infty} \prod_i [d\vec{a}_i \, d\theta_i] \sum_{\{\vec{n}_i\}} e^{-\frac{T}{2}\sum_i (\Delta \times \vec{a}_i)^2 - \frac{1}{2t}\sum_{i,\mu}(\theta_{i+\hat{\mu}} - \theta_i - 2\pi a_{i,\mu} - 2\pi n_{i,\mu})^2},$$
$$\tag{7.6}$$

which may be recognized as the Villain approximation of the *lattice super-conductor*:

$$Z_V = \lim_{t \to 0} Z_{LS}, \tag{7.7}$$

with

$$Z_{LS} = \int_{-\infty}^{\infty} \prod_i [d\vec{a}_i \, d\theta_i] e^{-\frac{T}{2}\sum_i (\Delta \times \vec{a}_i)^2 + \frac{1}{t}\sum_{i,\mu} \cos(\theta_{i+\hat{\mu}} - \theta_i - 2\pi a_{i,\mu})}. \tag{7.8}$$

The last expression describes the lattice version of the Ginzburg–Landau theory in Eq. (4.1) for the superconducting phase transition in the extreme type-II limit, $b \to \infty$, $a(T) \to -\infty$, $b/|a(T)| = \text{const}$, when the fluctuations of the amplitude of the order parameter are completely suppressed. The chemical potential for vortex loops t in this representation plays the role of temperature, and we will refer to the limit $t \to 0$ as the *frozen lattice superconductor*. After rescaling of the gauge field as $\sqrt{T}\vec{a} \to \vec{a}$ we see that the temperature T in the

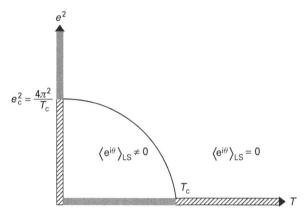

Figure 7.2 Phase diagram of the lattice superconductor at temperature T and with a charge e.

XY model determines the charge of the lattice superconductor as

$$e^2 = \frac{4\pi^2}{T}.$$ (7.9)

Since the original XY model may be viewed as a special case of a lattice superconductor with zero charge, the non-analytic part of the free energy for the lattice superconductor $F_{LS}(e^2, T)$ in three dimensions satisfies the *duality relation*

$$F_{LS}(0, T) = F_{LS}\left(\frac{4\pi^2}{T}, 0\right).$$ (7.10)

The left hand side of Eq. (7.10) refers to the XY model, and the right hand side to the frozen lattice superconductor.

As decreasing the temperature in the XY model increases the charge of the frozen lattice superconductor, the ordered phase of the XY model by duality maps onto the disordered phase of the frozen lattice superconductor, and vice versa. The phase diagram of the lattice superconductor is depicted in Fig. 7.2, with the regions exactly dual to each other marked alike. The $e^2 = 0$ axis represents the XY model, and the $T = 0$ axis represents the frozen lattice superconductor. Since a small temperature near the critical point of the frozen lattice superconductor should be irrelevant, and the XY model at T_c is exactly dual to the frozen lattice superconductor at $e^2 = 4\pi^2/T_c$, the whole transition line in the $T - e^2$ plane may be expected to be in the universality class of the frozen lattice superconductor. In particular, this implies that the specific heat of the lattice superconductor diverges with the XY exponent α_{XY}, but with the *inverted* temperature axis: the universal amplitude ratio defined in Eq. (1.24), for example, in the lattice superconductor should take the inverse of the XY value. Since the hyperscaling holds, the correlation length in the frozen lattice superconductor should also diverge with the XY exponent $\nu_{XY} = 0.67$. The other exponents, however, do not follow simply from the duality relation for the free energy, and are not all equal. The validity of the above reasoning and of the inverted XY critical behavior of the lattice superconductor has been confirmed in detailed numerical simulations. Since the lattice superconductor should have the phase transition in the same universality class as the type-II Ginzburg–Landau theory for a single complex field in three dimensions, this suggests that the flow diagram for the three-dimensional superconductor should be as in Fig. 4.9b, with the critical point C lying in the inverted XY universality class.

Problem 7.1 Show that the partition function

$$ Z = \int_0^{2\pi} \prod_i \frac{d\psi_i}{2\pi} \sum_{\{\vec{n}\}} e^{-\frac{T}{2}\sum_i (\Delta\times\vec{n})^2 - \frac{e^2}{8\pi^2}\sum_{i,\hat{\mu}}(\psi_i - \psi_{i-\hat{\mu}} - 2\pi n_{i,\mu})^2} $$

is dual to the Villain lattice superconductor with charge e and temperature T.

Solution Starting from the lattice superconductor in Villain's approximation,

$$ Z_{LS} = \int_{-\infty}^{\infty} \prod_i [d\vec{a}_i d\theta_i] \sum_{\{\vec{n}_i\}} e^{-\frac{1}{2e^2}\sum_i (\Delta\times\vec{a}_i)^2 - \frac{1}{2T}\sum_i (\theta_{i+\hat{\mu}} - \theta_i - a_{i,\mu} - 2\pi n_{i,\mu})^2}, $$

the introduction of two Hubbard–Stratonovich variables leads to

$$ Z_{LS} = \int_{-\infty}^{\infty} \prod_i [d\vec{a}_i d\theta_i d\vec{b}_i d\vec{X}_i] $$
$$ \times \sum_{\{\vec{n}_i\}} e^{-\frac{e^2}{2}\sum_i \vec{b}_i^2 + i\sum_i \vec{a}_i\cdot(\Delta\times\vec{b}_i) + i\sum_{i,\mu} X_{i,\mu}(\theta_{i+\hat{\mu}} - \theta_i - a_{i,\mu} - 2\pi n_{i,\mu}) - \frac{T}{2}\sum_i \vec{X}_i^2}. $$

Performing the sum and integrating over the phases and the gauge field yields $\vec{X}_i = \vec{m}_i$, $\Delta \cdot \vec{m}_i = 0$, with $\{\vec{m}\}$ integer, and $\Delta \times \vec{b}_i = \vec{m}_i$. We can therefore write $\vec{m}_i = \Delta \times \vec{n}_i$, so that $\vec{b} = \vec{n} - \Delta\psi_i/(2\pi)$, with ψ_i real. Since the integrand is invariant under the shift of all variables ψ_i by 2π, the partition function may be finally written as in the statement of the problem. Note that in the limits $T = 0$, or $e = 0$, the partition function Z reduces to the XY model.

7.2 Confinement of magnetic monopoles

Duality relations analogous to Eq. (7.10) also exist between certain correlation functions. Consider the dipole–dipole correlation function in the original XY model:

$$\langle e^{i\phi_N} e^{-i\phi_{N'}}\rangle_{XY} = Z_{XY}^{-1} \int_0^{2\pi} \prod_i \frac{d\phi_i}{2\pi} e^{\frac{1}{T}\sum_{i,\mu}\cos(\phi_i-\phi_{i-\hat{\mu}})+i(\phi_N-\phi_{N'})}. \quad (7.11)$$

In the Villain approximation, using the same set of transformations as in Eqs. (6.29)–(6.33), we find

$$\langle e^{i\phi_N} e^{-i\phi_{N'}}\rangle_{XY} = Z_{XY}^{-1} \int_0^{2\pi} \left(\prod_i \frac{d\phi_i}{2\pi}\right)$$

$$\times \sum_{\{n_{i,\mu}\}} e^{-\frac{T}{2}\sum_{i,\mu} n_{i,\mu}^2 -i\sum_{i,\mu} n_{i,\mu}(\phi_i-\phi_{i+\hat{\mu}})+i(\phi_N-\phi_{N'})}. \quad (7.12)$$

The integration over the phases now makes the integers $\{\vec{n}_i\}$ divergenceless everywhere, except at the sites N and N':

$$\sum_\mu (n_{i,\mu} - n_{i-\hat{\mu},\mu}) = \delta_{iN} - \delta_{iN'}. \quad (7.13)$$

We may therefore write

$$\vec{n}_i = \Delta \times \vec{m}_i + \vec{L}_i, \quad (7.14)$$

where the integer-valued vector $\vec{L}_i$ is finite and equal to ± 1 along an arbitrary path on the lattice that leads from the site N to N' (Fig. 7.3). Continuing with the duality transformations as in Eqs. (7.1)–(7.8), we obtain

$$\langle e^{i\phi_N} e^{-i\phi_{N'}}\rangle_{XY} = e^{-(F_{LS}(4\pi^2/T,0,\vec{L})-F_{LS}(4\pi^2/T,0,0))}, \quad (7.15)$$

where $F_{LS}(4\pi^2/T, 0, \vec{L})$ is the free energy of the frozen lattice superconductor in the presence of the vector $\vec{L}$,

$$F_{LS}(e^2, t, \vec{L}) = -\ln \int_{-\infty}^{\infty} \prod_i [d\vec{a}_i d\theta_i] e^{-\frac{1}{2}\sum_i(\Delta\times\vec{a}_i+\sqrt{T}\vec{L}_i)^2 -\frac{1}{t}\sum_i \cos(\theta_{i+\hat{\mu}}-\theta_i-ea_{i,\mu})}.$$

$$(7.16)$$

The configuration described by $\vec{L}_i$ is the *Dirac string* of magnetic flux between the magnetic monopoles located at N and N'. The relation between the magnetic charge $g = \sqrt{T}$ and the electric charge $e = 2\pi/\sqrt{T}$ is also known as the *Dirac quantization condition*:

$$ge = 2\pi. \quad (7.17)$$

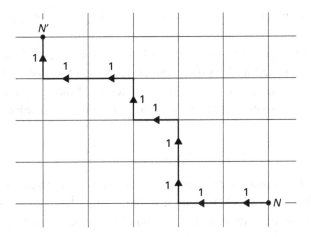

Figure 7.3 An example of the Dirac string $\vec{L}_i$ between the monopole at the site N and the antimonopole at the site N'.

It is possible now to use the knowledge of the behavior of the correlation function of the XY model to learn about the free energy of the Dirac string. At temperatures $T < T_c$ the XY model has long-range order. At very low temperatures we may use the Gaussian approximation, which gives

$$\langle e^{i\phi_N} e^{-i\phi_{N'}} \rangle_{XY} \approx \exp\left(-T \int \frac{d^3\vec{k}}{(2\pi)^3} \frac{1 - e^{i\vec{k}\cdot(\vec{N}-\vec{N'})}}{k^2} \right). \qquad (7.18)$$

Performing the integral and using the duality relation in Eq. (7.15), one finds

$$\Delta F_{LS}[\vec{L}] = F_{LS}(4\pi^2/T, 0, \vec{L}) - F_{LS}(4\pi^2/T, 0, 0) = \frac{T\Lambda}{8\pi^2} - \frac{T}{4\pi|N - N'|}, \qquad (7.19)$$

in the disordered phase of the frozen lattice superconductor, for $e > e_c = 2\pi/\sqrt{T_c}$. To create a widely separated monopole–antimonopole pair in the disordered phase of the superconductor costs a finite energy, and a monopole and an antimonopole attract via Coulomb interaction.

At high temperatures $T > T_c$, on the other hand, the correlation function in the XY model is short ranged:

$$\langle e^{i\phi_N} e^{-i\phi_{N'}} \rangle_{XY} \sim e^{-\frac{|N-N'|}{\xi(T)}}, \qquad (7.20)$$

with $\xi(T)$ as a finite correlation length. Duality implies then that in the ordered phase of the superconductor, for $e < e_c$,

$$\Delta F_{LS}[\vec{L}] \sim \frac{|N - N'|}{\xi(e)}, \qquad (7.21)$$

where the coefficient $1/\xi(e)$ vanishes as $e \to e_c$. Monopoles are therefore *confined* in the ordered phase of the superconductor, as the energy of the monopole–antimonopole pair increases linearly with the pair's size. This is just what is expected in the superconducting phase of the type-II superconductor where the magnetic flux is confined to an Abrikosov's flux tube that connects the monopoles. Right at $e = e_c$ the power law correlations in the XY model imply that

$$\Delta F_{LS} = (1 + \eta_{XY}) \ln |N - N'|, \qquad (7.22)$$

and right at the critical point monopoles interact via a logarithmic potential. The three-dimensional XY transition in the dual picture becomes therefore the confinement–deconfinement transition of the magnetic flux in the lattice superconductor.

Problem 7.2 Show that

$$\langle e^{i\theta_N} e^{-i\theta_{N'}} \rangle_{FLS} = e^{-\Delta F_{XY}(T, \vec{L})},$$

where $\Delta F_{XY}(T, \vec{L})$ is the excess of the free energy of the XY model in the presence of the vortex line $\vec{L}$ between the sites N and N'.

Solution Using the definition of the correlation function and reversing the duality transformations, the above correlator becomes the ratio of the expressions in Eq. (7.3), with $\Delta \cdot \vec{M}_i = 0$ in the denominator, and $\Delta \cdot \vec{M}_i = \delta_{iN} - \delta_{iN'}$ in the numerator. This implies the above duality relation.

Below T_c the free energy of a vortex line in the XY model depends linearly on its length, whereas above T_c it approaches a finite constant. The XY phase transition may therefore be identified with the point where the vortex line tension, defined as the ratio of the energy and the length of a long vortex line, vanishes. The low temperature phase consists therefore of fluctuating vortex loops of finite size which diverges at T_c. This is the three-dimensional analog of the Kosterlitz–Thouless transition. In particular, the above duality relation implies that $\langle e^{i\theta_i} \rangle_{LS}$ may be taken as the order parameter for the vortex transition, dual to the original order parameter $\langle e^{i\phi_i} \rangle_{XY}$.

7.3 Magnetic field correlations

It is easy to derive an additional duality relation between the current–current correlation function in the XY model and the magnetic field correlation function in the lattice superconductor. To this purpose let us introduce the source

$\vec{A}_i$ into the XY model in Eq. (6.29) by replacing

$$\phi_{i+\hat{\mu}} - \phi_i \rightarrow \phi_{i+\hat{\mu}} - \phi_i - A_{i,\mu}. \tag{7.23}$$

The current–current correlation function is then defined as

$$\langle J_{N,\mu} J_{N',\nu} \rangle_{XY} = -\frac{\delta^2}{\delta A_{N,\mu} \delta A_{N',\nu}} \ln Z_{XY}[\vec{A}]|_{\vec{A}=0}. \tag{7.24}$$

Retaining the source field during the duality transformations and differentiating at the end then yields

$$\langle J_{N,\mu} J_{N',\nu} \rangle_{XY} = \langle (\Delta \times \vec{a})_{N,\mu} (\Delta \times \vec{a})_{N',\nu} \rangle_{LS}. \tag{7.25}$$

In the disordered phase of the lattice superconductor the gauge field is massless and its dynamics is given by the standard Maxwell term. By duality this implies that

$$\langle J_\mu(\vec{q}) J_\nu(-\vec{q}) \rangle_{XY} = \frac{K}{T} (\delta_{\mu\nu} - \hat{q}_\mu \hat{q}_\nu) + O(q^2), \tag{7.26}$$

where we have turned to Fourier space for simplicity. The coefficient K in the above current–current correlator defines the *helicity modulus*, or the *phase stiffness* in the XY model, which is thus finite below T_c, and zero above it. For $T > T_c$, on the other hand, all correlation functions of the local variables in the XY model such as the current defined above must be short ranged. By duality then

$$\langle (\Delta \times \vec{a})_{N,\mu} (\Delta \times \vec{a})_{N',\nu} \rangle_{FLS} \sim e^{-\frac{|N-N'|}{\xi_{xy}(T)}} \tag{7.27}$$

in the lattice superconductor for $e < e_c$. This is just the familiar Meissner effect in the superconducting phase. The magnetic field penetration depth in the lattice superconductor is therefore

$$\lambda_{LS} \sim \xi_{xy}. \tag{7.28}$$

Since $\xi_{LS} \sim \xi_{xy}$ from the duality between the free energies and hyperscaling, it follows that

$$\lambda_{LS} \sim \xi_{LS}, \tag{7.29}$$

in agreement with Eq. (4.37), derived rather differently in Chapter 4.

Problem 7.3 Show that the current–current correlation function in the frozen lattice superconductor by duality maps onto the vortex–vortex correlation function in the XY model.

Solution By applying the shift in Eq. (7.23) to $\theta_{i+\hat{\mu}} - \theta_i$ in Eq. (7.8), reversing the duality transformations, and differentiating twice with respect to the source, one finds that

$$\langle J_{N,\mu} J_{N',\nu} \rangle_{\mathrm{FLS}} = \langle M_{N,\mu} M_{N',\nu} \rangle_{\mathrm{XY}}.$$

Problem 7.4 Determine the long-distance behavior of the current–current correlation function of the three dimensional XY model at the critical point.

Solution At the critical point of the lattice superconductor and in Fourier space:

$$\langle b_\mu(\vec{q}) b_\nu(-\vec{q}) \rangle_{\mathrm{LS}} \sim \frac{q^2}{q^{2-\eta_{\mathrm{A}}}} (\delta_{\mu\nu} - \hat{q}_\mu \hat{q}_\nu),$$

where $\vec{b} = \Delta \times \vec{a}$, and $\eta_{\mathrm{A}} = 1$ in three dimensions (Section 4.4). Duality implies then that

$$\langle J_{N,\mu} J_{N',\nu} \rangle_{\mathrm{XY}} \sim \frac{1}{|N - N'|^4},$$

at $T = T_{\mathrm{c}}$.

7.4 Compact electrodynamics*

As the final application of duality transformations, let us define the *compact quantum electrodynamics* by the partition function

$$Z_{\mathrm{CQED}} = \int_0^{2\pi} \prod_i d\vec{A}_i e^{g^{-2} \sum_{i,\mu\nu} \cos(F_{i,\mu\nu})}, \tag{7.30}$$

where $F_{i,\mu\nu}$ is the electromagnetic field tensor

$$F_{i,\mu\nu} = \Delta_\mu A_{i,\nu} - \Delta_\nu A_{i,\mu}. \tag{7.31}$$

Although the gauge field is not coupled to matter, Z_{CQED} may still have a singularity at some g_{c} due to periodicity of the action, in analogy with the XY model. In this section we focus on four dimensions, $\mu = 1, 2, 3, 4$, and show that the above Z_{CQED} is then dual to the four-dimensional frozen lattice superconductor. We will argue that the singularity at $g = g_{\mathrm{c}}$ corresponds to the transition from the phase with the massless photon at weak coupling to the one without it at strong coupling. Also an interesting three-dimensional example of compact electrodynamics is relegated to Problem 7.5.

As usual, in the Villain approximation Z_{CQED} is written as

$$Z_{CQED} = \int_0^{2\pi} \prod_i d\vec{A} \sum_{\{m_{i,\mu\nu}\}} e^{-\frac{1}{2g^2} \sum_{i,\mu\nu} (F_{i,\mu\nu} - 2\pi m_{i,\mu\nu})^2}. \tag{7.32}$$

Introducing Hubbard–Stratonovich variables $\{b_{i,\mu\nu}\}$ to decouple the quadratic term, and then integrating over the gauge field leads to

$$Z_{CQED} = \int \prod_{i,\mu\nu} db_{i,\mu\nu} \sum_{\{m_{i,\mu\nu}\}} \delta(\Delta_\mu b_{i,\mu\nu}) e^{-\frac{g^2}{2} \sum_{i,\mu\nu} b_{i,\mu\nu}^2 - i2\pi \sum_{i,\mu\nu} b_{i,\mu\nu} m_{i,\mu\nu}}. \tag{7.33}$$

The constraint may be resolved by writing

$$b_{i,\mu\nu} = \sum_{\lambda\sigma} \epsilon_{\mu\nu\lambda\sigma} f_{i,\lambda\sigma}, \tag{7.34}$$

with the antisymmetric dual field tensor $f_{i,\mu\nu}$ as

$$f_{i,\mu\nu} = \Delta_\mu a_{i,\nu} - \Delta_\nu a_{i,\mu}. \tag{7.35}$$

Using the identity

$$\sum_{\mu\nu} \epsilon_{\mu\nu\lambda\sigma} \epsilon_{\mu\nu\alpha\beta} = 2(\delta_{\lambda\alpha}\delta_{\sigma\beta} - \delta_{\lambda\beta}\delta_{\sigma\alpha}), \tag{7.36}$$

Z_{CQED} may then be written as

$$Z_{CQED} = \int \prod_{i,\mu} da_{i,\mu} \sum_{\{m_{i,\mu\nu}\}} e^{-2g^2 \sum_{i,\mu\nu} f_{i,\mu\nu}^2 - i2\pi \sum_{i,\mu\nu\lambda\sigma} \epsilon_{\mu\nu\lambda\sigma} f_{i,\lambda\sigma} m_{i,\mu\nu}}. \tag{7.37}$$

The last term in the exponent, on the other hand, can be written as

$$-i2\pi \sum_{i\sigma} a_{i\sigma} M_{i\sigma}, \tag{7.38}$$

with

$$M_{i,\sigma} = 2 \sum_{\mu\nu\lambda} \epsilon_{\mu\nu\lambda\sigma} \Delta_\lambda m_{i,\mu\nu}. \tag{7.39}$$

Since evidently

$$\sum_\sigma \Delta_\sigma M_{i,\sigma} = 0, \tag{7.40}$$

the four-component vectors $\{M_{i,\sigma}\}$ form closed loops on the four-dimensional lattice. Repeating the steps between Eqs. (7.4) and (7.8), one can rewrite

Eq. (7.37) as

$$Z_{\text{CQED}} = \lim_{t \to 0} \int \prod_{i,\mu} \mathrm{d}a_{i\mu} \mathrm{d}\theta_i \, \mathrm{e}^{\frac{1}{t} \sum_{i,\mu} \cos(\theta_{i+\hat{\mu}} - \theta_i - 2\pi a_{i,\mu}) - 2g^2 \sum_{i,\alpha\beta} f_{i,\alpha\beta}^2}, \quad (7.41)$$

which is the four-dimensional version of the partition function of the frozen lattice superconductor in Eq. (7.8). The charge of the superconductor is now

$$e = \frac{2\pi}{g}. \quad (7.42)$$

The weak coupling $(g < g_c)$ phase of the compact quantum electrodynamics by duality therefore maps onto the disordered $(e > e_c)$ phase of the lattice superconductor, and vice versa. The duality relation derived in Problem 7.2 implies that in the weak-coupling phase all closed loops of $\{M_{i,\sigma}\}$ are of finite size, whereas in the strong-coupling phase loops proliferate through the system, due to vanishing of the line tension. The point where the infinitely large loops first appear determines the critical coupling g_c. Based on the ϵ-expansion for the Ginzburg–Landau theory of the superconducting transition and the duality in four dimensions, one expects that the compact quantum electrodynamics has a weak fluctuation-induced first-order phase transition at $g = g_c$.

Further insight into the nature of the two phases of Z_{CQED} may be gained by deriving the equivalent of Eq. (7.25) in four dimensions. In analogy to Eq. (7.23), let us introduce the sources $G_{i,\mu\nu}$ into Eq. (7.32) as

$$F_{i,\mu\nu} \to F_{i,\mu\nu} - G_{i,\mu\nu}. \quad (7.43)$$

The current–current correlation function is then defined as

$$\langle J_{N,\mu\nu} J_{N',\lambda\sigma} \rangle_{\text{CQED}} = -\frac{\partial^2}{\partial G_{N,\mu\nu} \partial G_{N',\lambda\sigma}} \ln Z[G]|_{G=0}. \quad (7.44)$$

It readily follows that

$$\langle J_{N,\mu\nu} J_{N',\lambda\sigma} \rangle_{\text{CQED}} = \sum_{\alpha\beta\gamma\delta} \langle \epsilon_{\mu\nu\alpha\beta} f_{N,\alpha\beta} \epsilon_{\lambda\sigma\gamma\delta} f_{N',\gamma\delta} \rangle_{\text{LS}}. \quad (7.45)$$

In the disordered phase of the lattice superconductor the magnetic field correlator on the right hand side approaches a constant at small wavevectors, in analogy to Eq. (7.26). The above duality relation implies then the same for the current–current correlator in the weak-coupling phase of the compact electrodynamics. In analogy to the XY model, this means that there is a finite helicity modulus in this phase, or that the periodicity of Z_{CQED} is not important at long length scales for $g < g_c$. This is tantamount to all closed loops of $\{M_{i,\sigma}\}$

being of finite size at weak coupling. Dropping all finite integers $\{m_{i,\mu\nu}\}$ in Eq. (7.32) then reduces Z_{CQED} to ordinary electrodynamics with the massless photon excitation. In the ordered phase of the lattice superconductor, on the other hand, the dual gauge field $a_{i,\mu\nu}$ becomes massive due to the Higgs mechanism. Duality implies then that

$$\langle J_{\mu\nu}(\vec{q})J_{\lambda\sigma}(-\vec{q})\rangle_{\text{CQED}} \sim q^2, \qquad (7.46)$$

for $g > g_c$, which vanishes when $q \to 0$. This means that the photon that existed at weak coupling becomes massive in the strong coupling phase. This qualitative difference may be used to distinguish the two different phases of the compact electrodynamics.

Problem 7.5* Determine the phase diagram of the pure compact quantum electrodynamics in three dimensions.

Solution Using the identity

$$\sum_{\alpha} \epsilon_{\alpha\mu\nu}\epsilon_{\alpha\lambda\sigma} = \delta_{\mu\lambda}\delta_{\nu\sigma} - \delta_{\mu\sigma}\delta_{\nu\lambda},$$

in three dimensions the electromagnetic tensor in Eq. (7.31) may be written as

$$F_{i,\mu\nu} = \sum_{\alpha\lambda\sigma} \epsilon_{\alpha\mu\nu}\epsilon_{\alpha\lambda\sigma}\Delta_\lambda A_{i,\sigma}.$$

In the Villain approximation the partition function is again as in Eq. (7.33), but with the constraint

$$\sum_{\lambda\sigma\alpha} \epsilon_{\sigma\lambda\alpha}\epsilon_{\alpha\mu\nu}\Delta_\lambda b_{i,\mu\nu} = 0.$$

This means that

$$\sum_{\mu\nu} \epsilon_{\alpha\mu\nu}b_{i,\mu\nu} = \Delta_\alpha\Phi_i,$$

or, by recalling that $b_{i,\mu\nu}$ is an antisymmetric tensor,

$$b_{i,\mu\nu} = \frac{1}{2}\sum_{\alpha} \epsilon_{\alpha\mu\nu}\Delta_\alpha\Phi_i.$$

Inserting $b_{i,\mu\nu}$ into Eq. (7.33) and summing over integers $\{m_{i,\mu\nu}\}$ yields

$$Z_{\text{CQED}} = \sum_{\{n\}} e^{-\frac{g^2}{4}\sum_{i,\alpha}(\Delta_\alpha n_i)^2}$$

in three dimensions.

The final result for the partition function is thus equivalent to Eq. (6.42), with $g^2 = 2T/J$ and in an additional dimension. Proceeding with the same transformations as in Eqs. (6.42)–(6.48) leads to sine-Gordon theory, which describes the neutral Coulomb plasma. In three dimensions, however, the Coulomb plasma is always in the metallic phase (Problem 6.4), and compact electrodynamics consequently has no phase transition. Since the current–current correlation function is

$$\langle J_{N,\mu\nu} J_{N',\lambda\sigma} \rangle_{\text{CQED}} = \frac{1}{4} \sum_{\alpha\beta} \epsilon_{\alpha\mu\nu} \epsilon_{\beta\lambda\sigma} \langle \Delta_\alpha \Phi_N \Delta_\beta \Phi_{N'} \rangle_{\text{sG}},$$

it follows that

$$\langle J_{\mu\nu}(q) J_{\lambda\sigma}(-q) \rangle_{\text{CQED}} \sim \frac{q^2}{\Sigma}$$

with $\Sigma \neq 0$ (Problem 6.4). The photon is thus always massive and the compact electrodynamics has only the strongly coupled phase in three dimensions.

Problem 7.6* Derive the theory dual to the compact three-dimensional lattice superconductor, defined by the action

$$S = -K \sum_{i,\mu} \cos(\Delta_\mu \theta_i - q a_{i,\mu}) - \beta \sum_{i,\mu\nu} \cos F_{i,\mu\nu}.$$

Solution In the Villain approximation, using two Hubbard–Stratonovich variables $b_{i,\mu}$ and $c_{i,\mu}$ one finds

$$S = \frac{1}{2K} \sum_{i,\mu} b_{i,\mu}^2 + i \sum_{i,\mu} b_{i,\mu} (\Delta_\mu \theta_i - q a_{i,\mu} - 2\pi m_{i,\mu})$$
$$+ \frac{1}{2\beta} \sum_{i,\mu\nu} c_{i,\mu\nu}^2 + i \sum_{i,\mu\nu} c_{i,\mu\nu} \left(\left(\sum_{\sigma\rho\beta} \epsilon_{\mu\nu\beta} \epsilon_{\beta\rho\sigma} \Delta_\rho a_{i,\sigma} \right) - 2\pi n_{i,\mu\nu} \right).$$

The integration over $\{\theta_i\}$ and $\{a_{i,\mu}\}$ yields the constraints

$$\sum_\mu \Delta_\mu b_{i,\mu} = 0,$$

and

$$q b_{i,\sigma} + \sum_{\mu\nu\beta\rho} \epsilon_{\mu\nu\beta} \epsilon_{\beta\rho\sigma} \Delta_\rho c_{i,\mu\nu} = 0,$$

whereas the summations over the integers $\{m_{i,\mu}\}$ and $\{n_{i,\mu\nu}\}$ force the Hubbard–Stratonovich variables $\{b_{i,\mu}\}$ and $\{c_{i,\mu\nu}\}$ to be integer as well. We can therefore write

$$b_{i,\sigma} = \sum_{\beta\rho} \epsilon_{\sigma\rho\beta} \Delta_\rho k_{i,\beta},$$

with $\{k_{i,\beta}\}$ integer, to satisfy the first constraint, and then

$$qk_{i,\beta} - \sum_{\mu\nu} \epsilon_{\beta\mu\nu} c_{i,\mu\nu} = \Delta_\beta M_i,$$

with $c_{i,\mu\nu}$ and M_i integers, to satisfy the second constraint. It follows then:

$$2c_{i,\mu\nu} = \sum_\beta (q\epsilon_{\mu\nu\beta} k_{i,\beta} - \epsilon_{\mu\nu\beta} \Delta_\beta M_i).$$

The action may therefore be written as

$$S = \frac{1}{2K} \sum_i (\Delta \times \vec{k}_i)^2 + \frac{1}{4\beta} \sum_{i,\mu} (\Delta_\mu M_i - qk_{i,\mu})^2,$$

where all the variables are integers.

Problem 7.7* Show that the partition function of the compact lattice super-conductor from the previous problem may also be written as

$$S = \frac{1}{2K} \sum_i (\Delta \times \vec{\Psi}_i)^2 + \frac{q^2}{4\beta} \sum_i \vec{\Psi}_i^2 + i2\pi \sum_i \vec{\Psi}_i \cdot \vec{M}_i,$$

with $\{\vec{\Psi}_i\}$ real, and $\Delta \cdot \vec{M}_i = qN_i$, with $\{\vec{M}_i\}$ and $\{N_i\}$ integer. Deduce the phase diagram when $q = 1$.

Solution From the solution of the previous problem, we may write

$$S = \frac{1}{2K} \sum_i (\Delta \times \vec{\phi}_i)^2 + \frac{q^2}{4\beta} \sum_{i,\mu} \left(\frac{\Delta_\mu a_i}{q} - \phi_{i,\mu}\right)^2 + i2\pi \sum_i (a_i N_i + \vec{\phi}_i \cdot \vec{M}_i),$$

where only $\{N_i\}$ and $\{\vec{M}_i\}$ are integers. Introducing $\vec{\Psi}_i = \vec{\phi}_i - \Delta(a_i/q)$ and then integrating over $\{a_i\}$ yields the above representation of the partition function.

For $q = 1$, the summation over integers $\{N_i\}$ makes the sum over $\{\vec{M}_i\}$ unconstrained. We may then proceed exactly as in the derivation of the sine-Gordon theory, and approximate S with

$$S = \frac{1}{2K} \sum_i (\Delta \times \vec{\Psi}_i)^2 + \frac{1}{4\beta} \sum_{i,\mu} \Psi_{i,\mu}^2 - 2y \sum_{i,\mu} \cos(2\pi \Psi_{i,\mu}),$$

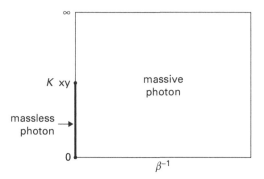

Figure 7.4 Conjectured phase diagram of compact three-dimensional lattice superconductor with $q = 1$.

with small fugacity y. The last action may be interpreted as describing a three-dimensional plasma of three-flavor charges interacting via a short-range interaction. Since by the arguments in Section 6 such a system should always be in the metallic phase, the original compact lattice superconductor is in the phase with a massive photon at any finite coupling β^{-1}. The phase diagram should therefore be as given in Fig. 7.4.

Problem 7.8* Determine the phases of the compact three-dimensional lattice superconductor from the previous two problems in the limit $K = \infty$ for both $q = 1$ and $q = 2$.

Solution When $K = \infty$, only the configurations where

$$\Delta_\mu \theta_i - q a_{i,\mu} = 2\pi n_{i,\mu}$$

with $\{n_{i,\mu}\}$ integer give a finite contribution to the partition function. The action for these is

$$S = -\beta \sum_{i,\mu\nu} \cos\left(\frac{2\pi}{q} \sum_{\beta\rho\sigma} \epsilon_{\mu\nu\beta} \epsilon_{\beta\rho\sigma} \Delta_\rho n_\sigma\right).$$

When $q = 1$ the action in units of β is thus simply a constant, and there are no phase transitions, in accord with the phase diagram in Fig. 7.4. When $q = 2$, however, we may write

$$S = -\beta \sum_{i,\mu\nu} \cos(\pi(\Delta_\mu n_{i,\nu} - \Delta_\nu n_{i,\mu})) = -2\beta \sum_{i,\alpha} \cos(\pi m_{i,\alpha}),$$

where $m_{i,\alpha} = \sum_{\mu\nu} \epsilon_{\alpha\mu\nu} \Delta_\mu n_{i,\nu}$ are also integer. The last expression is known as the Z_2 gauge theory, and it is dual to the Ising model. To see this, set $K = \infty$ in the final result of Problem 7.6. For $q = 2$ we may then write

$$S = \frac{1}{4\pi^2\beta} \sum_{i,\mu} (\pi \Delta_\mu M_i - 2\pi k_{i,\mu})^2,$$

which is the Villain approximation to

$$S = -\frac{1}{2\pi^2\beta} \sum_{i,\mu} \cos(\pi \Delta_\mu M_i).$$

Since each pair (i, μ) labels a link on the quadratic lattice, and each link according to the last expression contributes just a sign to the Boltzmann weight, we may recognize the partition function in this limit as equivalent to the Ising model. For $q = 2$ then, in contrast to when $q = 1$, there is a phase transition when $K = \infty$, and it is in the Ising universality class.

8

Quantum phase transitions

The dynamical critical exponent is introduced. The phase diagram and the phase transitions in the Bose–Hubbard model of interacting bosons on a lattice are determined. The concept of quantum fluctuations is introduced on the example of an interacting superfluid, and finally the special scaling of conductivity is discussed.

8.1 Dynamical critical exponent

The finite temperature phase transitions studied in previous chapters are the result of the competition between the entropy and the energy terms in the free energy: the weight of entropy increases with temperature, and ultimately destroys the order that may be existing in the system. A sharp phase transition between two phases exhibiting qualitatively different correlations may, however, occur even at zero temperature, by varying a coupling constant in the Hamiltonian. The transition then corresponds to a non-analyticity of the energy of the ground state. A simple example is provided by the interacting bosons, where the superfluid transition may be brought about by tuning the chemical potential at $T = 0$. Such $T = 0$ phase transitions are called *quantum phase transitions*, and will be the subject of the present chapter.

In general, a quantum phase transition may lie at the end of a line of thermal phase transitions, as in the bosonic example mentioned above. It is possible, however, that the system may not even have an ordered state at finite temperatures, but still exhibits a quantum critical point. Two different situations are depicted in Fig. 8.1. Examples of such phase diagrams are provided by the system of interacting bosons which will be discussed in Sections 8.2 and 8.3.

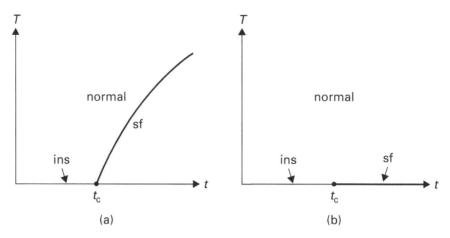

Figure 8.1 Phase diagram of interacting bosons at a density commensurate with the lattice in two (a) and one (b) dimension. The parameter t measures hopping between the nearest-neighboring sites. (a) The system has the Kosterlitz–Thouless transition at $T \neq 0$, and the three-dimensional XY quantum phase transition at $T = 0$. (b) The system exhibits the Kosterlitz–Thouless transition between the insulating and the superfluid phases at $T = 0$ and $t = t_c$, and no phase transition at $T \neq 0$.

As mentioned in the Section 2.3, the principal difference between the quantum ($T = 0$) and the thermal ($T \neq 0$) phase transitions is that while dynamics is irrelevant for the latter, it is crucial for the former. Put differently, whereas near a finite temperature transition the fluctuating order parameter may be considered completely independent of the imaginary time, this is no longer possible in the quantum case. Besides the usual fluctuations in space, the fluctuations in imaginary time need to be taken into account as well, and thus the imaginary time effectively plays the role of an additional dimension of the system. As the system's dimensionality is the crucial factor in determining its critical behavior, it is obviously important to determine how exactly the imaginary time should be counted relative to spatial dimensions. This brings us to the notion of an additional critical exponent at quantum phase transitions, called the *dynamical critical exponent*.

Let us reconsider the superfluid susceptibility in Eq. (3.1). At $T = 0$ it will in principle be a function of the wavevector $\vec{k}$, the chemical potential μ, the interaction λ, and the Matsubara frequency ω, i.e.

$$\chi(\vec{k}, \omega) = F(k, \omega, \mu, \lambda). \tag{8.1}$$

The susceptibility is, of course, still implicitly dependent on the ultraviolet cutoff Λ, and renormalizability implies that

$$\chi(\vec{k}, \omega) = b^x F(bk, b^z \omega, \mu(b), \lambda(b)),$$ (8.2)

in analogy with Eq. (3.4). The power x and the functions $\mu(b)$ and $\lambda(b)$ will in principle be different than at $T \neq 0$. Most importantly, the Matsubara frequency and the wavevector may be rescaled differently. The exponent z that quantifies this difference, and which may or may not be unity, is the dynamical critical exponent.

Assume that for $b \gg 1$, $\mu(b) \approx \mu b^{1/\nu}$ and $\lambda(b) \approx \lambda^*$, where ν and λ^* characterize the quantum critical point in the system. Choosing again the parameter b so that $\mu b^{1/\nu} = \mu_0$ with μ_0 constant, the susceptibility at $T = 0$ may be written as

$$\chi(\vec{k}, \omega) = \left| \frac{\mu_0}{\mu} \right|^{x\nu} F(k\xi, \omega\tau, \mu_0, \lambda^*),$$ (8.3)

where $\xi = |\mu_0/\mu|^\nu$ is the correlation length, and $\tau = \xi^z$ the *correlation "time"*. The dynamical exponent therefore may be understood as connecting the characteristic time scale and length scale for fluctuations near the quantum critical point.

The reader may be wondering about the relevance of the exponent z in experiments, which, of course, are always performed at finite temperatures. Consider therefore the finite temperature susceptibility, satisfying a similar scaling equation to Eq. (8.2):

$$\chi(\vec{k}, \omega) = b^x F(bk, b^z \omega, T(b), \mu(b), \lambda(b)).$$ (8.4)

As the inverse of temperature enters only as the upper limit of integration in imaginary time, temperature must be rescaled exactly inversely to imaginary time. This therefore is the same scaling as the frequency. So, $T(b) = b^z T$, and the dynamical critical exponent also governs the finite temperature scaling near the quantum critical point. In particular, the uniform static susceptibility is then

$$\chi(0, 0) = \left| \frac{\mu_0}{\mu} \right|^{x\nu} F\left(0, 0, T \left| \frac{\mu_0}{\mu} \right|^{z\nu}, \mu_0, \lambda^* \right),$$ (8.5)

and the critical temperature is therefore

$$T_c = c \left(\frac{\mu}{\mu_0} \right)^{z\nu},$$ (8.6)

with c as a non-universal constant with units of temperature. The dynamical critical exponent therefore directly determines the form of the phase boundary in the $\mu - T$ plane near the quantum critical point at $T = 0$. Right at the quantum critical point at $\mu = 0$, on the other hand, we may choose $b^z T = T_0$, where T_0 is some fixed temperature. Then

$$\chi(\vec{k}, \omega) = \left(\frac{T_0}{T}\right)^{x/z} F\left(k\left(\frac{T_0}{T}\right)^{\frac{1}{z}}, \omega\left(\frac{T_0}{T}\right), T_0, 0, \lambda^*\right), \quad (8.7)$$

and the temperature itself provides the characteristic length and time scales for fluctuations.

Since the effective dimensionality of the system at $T = 0$ is always increased by the addition of the imaginary time direction, it often happens that in physical dimensions quantum critical behavior becomes trivial, although the thermal critical behavior at the same transition was not. The reader may thus legitimately worry whether, this being the case, quantum phase transitions require a separate chapter in a book on critical phenomena. We believe the answer is yes, nevertheless, for at least two reasons. First, when the quantum critical behavior is non-trivial, quantum phase transitions often are qualitatively different from the thermal phase transitions, due to the special role played by the imaginary time. Examples of this will be provided in the subsequent sections. Second, in the experiments one is of course interested in the dynamics of the system in real and not imaginary time. This in principle is computable from the response known in the imaginary time, or at the Matsubara frequencies, by analytically continuing to real time, or equivalently to real frequencies. The step of analytical continuation, however, may be highly non-trivial when one knows the function to be continued only approximately. Some, maybe deceivingly simple, examples of this will also be discussed in the subsequent sections.

8.2 Quantum critical point in Φ^4-theory

As a paradigmatic example of a quantum phase transition let us consider a general Φ^4 action given by

$$S = \int_0^{1/T} d\tau \int d\vec{r}[u\Phi^*(\vec{r}, \tau)\partial_\tau \Phi(\vec{r}, \tau) + v|\partial_\tau \Phi(\vec{r}, \tau)|^2$$
$$+ |\nabla \Phi(\vec{r}, \tau)|^2 - \mu|\Phi(\vec{r}, \tau)|^2 + \lambda|\Phi(\vec{r}, \tau)|^4]. \quad (8.8)$$

To be specific we assumed that $\Phi(\vec{r}, \tau)$ is complex, but an analogous action may be written for a field with an arbitrary number of real components, and

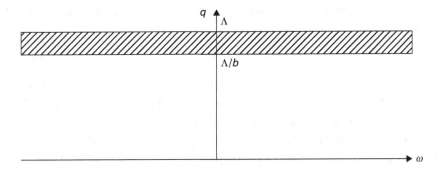

Figure 8.2 The shaded region represents the high-energy modes that are integrated out in the renormalization procedure at $T = 0$.

in particular for the Ising and the Heisenberg universality classes. The τ-dependence of Φ may be neglected at the finite temperature transition, and the universality class is then as described in Chapter 2. At $T = 0$, however, the first two terms must be taken into account, and as a result the universality class of the quantum phase transition will be different from that of its thermal counterpart.

When $u \neq 0$, $v = 0$, and Φ is complex, the above action describes the Bose–Einstein condensation at $T = 0$. For $u = 0$, $v \neq 0$, and a complex Φ, it describes the superfluid–Mott insulator transition in a system of lattice bosons at a commensurate density, as will be discussed in detail in the next section. When $u = 0$ and Φ has one or three real components, S describes the quantum phase transition between magnetically ordered and disordered states of the quantum versions of the Ising and the Heisenberg model, respectively.

Let us assume first that $u = 0$ and that Φ is complex, and determine the flow of the coupling constants at $T = 0$ under the change of cutoff as $\Lambda \rightarrow \Lambda/b$. Following exactly the same procedure as in Section 3.2, integrating the modes with $\Lambda/b < \vec{k} < \Lambda$ and $-\infty < \omega < \infty$, as depicted in Fig. 8.2, to the leading order in λ yields

$$\mu(b) = b^2(\mu - 4\lambda \int_{-\infty}^{\infty} \frac{d\omega}{2\pi} \int_{\Lambda/b}^{\Lambda} \frac{d\vec{q}}{(2\pi)^d} \left(\frac{1}{\omega^2 + k^2 - \mu} - \frac{1}{\omega^2 + k^2} \right), \quad (8.9)$$

$$\lambda(b) = b^{4-d-z}\lambda \left(1 - 10\lambda \int_{-\infty}^{\infty} \frac{d\omega}{2\pi} \int_{\Lambda/b}^{\Lambda} \frac{d\vec{q}}{(2\pi)^d} \frac{1}{(\omega^2 + k^2 - \mu)^2} \right), \quad (8.10)$$

where for convenience we have set $v = 1$. The dynamical exponents are determined by the requirement that after rescaling of momenta as $bk \rightarrow k$ and of frequencies as $b^z\omega \rightarrow \omega$ one can rescale the field Φ to bring the action into the old form. In the present case frequency and momentum appear symmetrically, and therefore

$$z = 1. \tag{8.11}$$

Defining then $\epsilon = 3 - d$, and the dimensionless coupling $\hat{\lambda} = \lambda \Lambda^{d-3} S_d/(4(2\pi)^d)$, the flow equations become identical to Eqs. (3.37) and (3.38), with the correlation length exponent given by Eq. (3.40) and $\eta = O(\epsilon^2)$.

One can show that the above result is completely general and not specific to the lowest order calculation nor to complex Φ. The quantum critical behavior in the above theory with $u = 0$ is identical to the classical critical behavior in the same theory in the dimension higher by one. The upper critical dimension for the transition at $T = 0$ is thus by one smaller than at $T \neq 0$.

The result is quite different when $u \neq 0$. First, in this case the coefficient v becomes irrelevant and may be set to zero. Second, to the first order in λ, only the diagram in Fig. 3.2 (b) yields a finite contribution. Setting the momenta and frequencies of the external legs to zero, the diagram in Fig. 3.2 (a), for example, gives

$$-8\lambda^2 \int_{-\infty}^{\infty} \frac{d\omega}{2\pi} \int_{\Lambda/b}^{\Lambda} \frac{d\vec{q}}{(2\pi)^d} \frac{1}{(i\omega + q^2 - \mu)(i\omega + q^2 - \mu)}, \tag{8.12}$$

where we have also set $u = 1$. Since both poles lie in the same complex half-plane, the integral over frequencies gives zero. The diagram 3.2 (b), on the other hand, gives, at $\mu = 0$,

$$-2\lambda^2 \int_{-\infty}^{\infty} \frac{d\omega}{2\pi} \int_{\Lambda/b}^{\Lambda} \frac{d\vec{q}}{(2\pi)^d} \frac{1}{(-i\omega + q^2)(i\omega + q^2)}$$

$$= -\frac{\lambda^2 S_d \Lambda^{d-2}}{(2\pi)^d(d-2)} \left(1 - \frac{1}{b^{d-2}}\right). \tag{8.13}$$

Defining a new small parameter as $\epsilon = 2 - d$, to the lowest order in ϵ the β-function for the interaction is

$$\beta_\lambda = \epsilon \hat{\lambda} - \hat{\lambda}^2 + O(\hat{\lambda}^3), \tag{8.14}$$

where $\hat{\lambda} = \lambda \Lambda^{d-2} S_d/(2\pi)^d$. As may already be suspected from the definition of ϵ, rescaling the momenta and the frequency now demands

$$z = 2, \tag{8.15}$$

and the upper critical dimension now equals two. The quantum critical behavior when $u \neq 0$ in both two and three dimensions is therefore the one of free bosons. This is in stark contrast with the critical behavior at finite temperatures, which as we saw lay in the universality class of the XY model. The

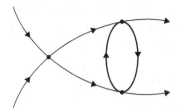

Figure 8.3 A vanishing third-order diagram in β_λ.

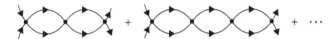

Figure 8.4 Higher-order diagrams included in the β_λ in Eq. (8.14).

reason for the triviality of the quantum criticality of interacting bosons is that the normal phase is actually an empty space without any particles. The quantum critical point therefore occurs at zero density of particles and is therefore not a many-body phenomenon at all.

It is interesting to note that Eq. (8.14) is an exact result, and there are actually no higher order terms in the β_λ. The reader may convince himself that a diagram in Fig. 8.3, for example, vanishes due to one of the frequency integrals. This is in fact true for all other higher order diagrams as well. The only exception are the ones in Fig. 8.4, which are proportional to $\lambda^n (\ln b)^{n-1}$, with $n \geq 3$. Being of higher than linear order in the small quantity $\ln b$, however, these do not yield any new terms in β_λ. They are already being included in the result in Eq. (8.14), as may be realized by computing $\hat{\lambda}(b)$ from the second order β-function.

Although the critical point for $d < 2$ is evidently non-Gaussian, the quantum critical behavior at the Bose–Einstein condensation is still trivial even in $d = 1$. This is because the usual diagrams in Figs. 3.1, 3.2 (c), and 3.2 (b) that would yield the lowest-order corrections to critical exponents, together with all the higher-order diagrams, vanish due to the frequency integration. So $\nu = 1/2$, and $\eta = 0$ exactly in all dimensions, including those below the upper critical dimension. This again simply reflects that the normal phase at $T = 0$ contains no particles.

Problem 8.1 Determine the thermal and the quantum critical behavior in the model defined by the action in Eq. (8.8), assuming that the field Φ has an infinite number of complex components.

Solution Generalizing the quartic term as in Problem 2.3 and utilizing the Hubbard–Stratonovich transformation, when the number of components $N \to \infty$ the free energy becomes

$$\frac{F}{NT} = (\chi(T) - \mu)|\sigma(T)|^2 - \frac{\chi^2(T)}{2\lambda}$$

$$+ T \sum_{\omega_n} \int \frac{d\vec{k}}{(2\pi)^d} \ln\left(i2u\omega_n + \omega_n^2 + k^2 + \chi(T) - \mu\right),$$

with $\chi(T)$ and $\sigma(T)$ satisfying the equations:

$$\frac{\chi(T)}{\lambda} = |\sigma(T)|^2 + T \sum_{\omega_n} \int \frac{d\vec{k}}{(2\pi)^d} \frac{1}{i2u\omega_n + \omega_n^2 + k^2 + \chi(T) - \mu}$$

and

$$(\chi(T) - \mu)\sigma(T) = 0.$$

Here we set the coefficient $v = 1$, $k_B = 1$, $u \to 2u$, and assumed a uniform Hubbard–Stratonovich field $\chi(T)$ at the saddle point. Matsubara frequencies are $\omega_n = 2\pi nT$, with n being a positive or negative integer, including zero.

In the disordered phase $\sigma(T) = 0$, and we may write

$$(\chi(T) - \mu)$$

$$\times \left[1 + \lambda T \sum_{\omega_n} \int \frac{d\vec{k}}{(2\pi)^d} \frac{1}{\left(i2u\omega_n + \omega_n^2 + k^2 + \chi(T) - \mu\right)\left(i2u\omega_n + \omega_n^2 + k^2\right)}\right]$$

$$= \mu_c(T) - \mu,$$

with the critical value of the chemical potential

$$\mu_c(T) = \lambda T \sum_{\omega_n} \int \frac{d\vec{k}}{(2\pi)^d} \frac{1}{i2u\omega_n + \omega_n^2 + k^2}.$$

At $T = 0$, the sum $T \sum_{\omega_n}$ may be replaced by an integral $\int d\omega/2\pi$, and for $u = 0$ the quantum critical behavior is the same as in Problem 2.3 with one extra dimension. For $u \neq 0$, performing the frequency integral at $T = 0$, we find

$$(\chi(0) - \mu)\left[1 + \frac{\lambda}{4}\int \frac{d\vec{k}}{(2\pi)^d} \frac{1}{(k^2 + u^2)^{3/2}} + O(\chi(0) - \mu)\right] = \mu_c(0) - \mu.$$

The integral over momenta is now always finite, and consequently the mean-field result

$$\chi(0) - \mu \propto \mu_c - \mu$$

obtains. At the quantum critical point $\gamma = 1$ and $\eta = 0$ in all dimensions, with the remaining critical exponent assuming their mean-field values as well.

At $T \neq 0$, on the other hand, we may separate the $\omega_n = 0$ term in the frequency sum. The rest is then always finite as $\chi(T) - \mu \to 0$, while the $\omega_n = 0$ term is finite for $d > 4$, and diverges as $(\chi(T) - \mu)^{(d-4)/2}$ for $2 < d < 4$. The thermal critical behavior in the theory is then the same as in Problem 2.3.

Finally, it is of interest to find the phase boundary in the $\mu - T$ plane. The sum over Matsubara frequencies yields

$$\mu_c(T) = \lambda \int \frac{d\vec{k}}{(2\pi)^d} \frac{n_B(u + \sqrt{u^2 + k^2}) - n_B(u - \sqrt{u^2 + k^2})}{2\sqrt{u^2 + k^2}},$$

where $n_B(z) = (e^{z/T} - 1)^{-1}$. We find:

$$\mu_c(T) - \mu_c(0) = \lambda \int \frac{d\vec{k}}{(2\pi)^d} \frac{n_B(u + \sqrt{u^2 + k^2}) + n_B(-u + \sqrt{u^2 + k^2})}{2\sqrt{u^2 + k^2}}.$$

At low temperatures $T \ll u$, therefore

$$\mu_c(T) - \mu_c(0) \propto \frac{\lambda}{2u} \int \frac{d\vec{k}}{(2\pi)^d} e^{-\frac{k^2}{2uT}} \sim T^{\frac{d}{2}},$$

in agreement with the result below Eq. (2.35). Similarly, for $u \ll T$ one finds $\mu_c(T) - \mu_c(0) \sim T^{d-1}$.

Problem 8.2 Assuming that the field Φ in the previous problem is charged, find the dependence of the magnetic field penetration depth on temperature.

Solution Let us introduce the electromagnetic potential by the usual substitution $\nabla \to \nabla - i\vec{A}(\vec{r})$, assuming that the charge is unity. Integrating out $N - 1$ components of the field, the free energy in the presence of the electromagnetic potential becomes

$$\frac{F[\vec{A}]}{NT} = \frac{F[0]}{NT} + \int d\vec{x}\, d\vec{y}\, \Pi_{ij}(\vec{x} - \vec{y}) A_i(\vec{x}) A_j(\vec{y}) + O(\vec{A}^4),$$

with the polarization

$$\Pi_{ij}(\vec{q}) = |\sigma(T)|^2 \delta_{ij} + \left(1 - \frac{1}{N}\right) \Pi'_{ij}(\vec{q}).$$

The response to the electromagnetic field consists of two parts: the first is provided by the condensate $|\sigma(T)|^2$, and the second by the remaining $N - 1$ non-condensed components. Since for $T < T_c$ the quadratic coefficient for those

$N - 1$ components is pinned at $\chi(T) - \mu = 0$, Π' is temperature independent below T_c. This guarantees that $\Pi'_{ij}(q) \to 0$ as $q \to 0$. The condensate, on the other hand, is

$$|\sigma(T)|^2 = \frac{\mu}{\lambda} - T \sum_{\omega_n} \int \frac{d\vec{k}}{(2\pi)^d} \frac{1}{i2u\omega_n + \omega_n^2 + k^2}.$$

The penetration depth is $\Lambda(T) = 1/|\sigma(T)|$, and

$$|\sigma(0)|^2 - |\sigma(T)|^2 = \frac{\mu_c(T) - \mu_c(0)}{\lambda},$$

with the right hand side as determined in the previous problem.

8.3 Bose–Hubbard model

Let us introduce the simplest model of interacting bosons that exhibits the transitions discussed in the previous section. Define the Hamiltonian

$$\hat{H}_{\text{BH}} = - \sum_{i,j} t_{ij} \hat{b}_i^\dagger \hat{b}_j - \mu \sum_{i=1}^N \hat{n}_i + \frac{U}{2} \sum_{i=1}^N \hat{n}_i(\hat{n}_i - 1), \qquad (8.16)$$

where i numerates the sites of a quadratic lattice, $\hat{n}_i = \hat{b}_i^\dagger \hat{b}_i$, and $\hat{b}_i^\dagger$ and $\hat{b}_i$ are the standard bosonic creation and annihilation operators satisfying the commutation relation $[\hat{b}_i, \hat{b}_j^\dagger] = \delta_{ij}$. This Hamiltonian describes a collection of bosons hopping on a lattice and interacting when on the same site with a repulsion $U > 0$. It is often called *the Bose–Hubbard model*, in analogy with its better known fermionic version.

The Bose–Hubbard model may be solved exactly when there is no hopping, and $t_{ij} \equiv 0$. Then $[\hat{H}_{\text{BH}}, \hat{n}_i] = 0$, and the ground state is simply

$$|0\rangle = \prod_i |n_i\rangle, \qquad (8.17)$$

where the numbers n_i are all equal to the smallest integer larger than μ/U: $n_i = n$, for $n - 1 < \mu/U < n$. The $N(N - 1)$ lowest excited states consist simply of removing a particle at one site and placing it at another. The ground state is non-degenerate, except when μ/U is integer, when it becomes 2^N-degenerate. If the bosons were carrying an electrical charge, the system at $t_{ij} = 0$ would be an insulator, with a finite charge gap in the spectrum. It is also *incompressible*, since the particle density is, within each interval for μ, independent of the chemical potential. Since the insulating behavior is

entirely due to the interaction, such an insulator is often referred to as the
Mott insulator.

Does the gapped insulating phase persist to finite values of hopping? With
$t_{ij} \neq 0$ the Bose–Hubbard model is no longer exactly solvable, but one may
still understand the qualitative features of its phase diagram using the mean-
field theory first, and then considering the fluctuations.

The equilibrium behavior of the Bose–Hubbard model follows from its
partition function, which, in complete analogy to Eq. (2.20), may again be
expressed as a coherent-state path integral:

$$Z = \int \prod_i \mathrm{D}b_i^*(\tau)\mathrm{D}b_i(\tau) \exp \left(\int_0^\beta \mathrm{d}\tau \left[\sum_i b_i^*(\tau)\partial_\tau b_i(\tau) - H_{\mathrm{BH}}(b^*(\tau), b(\tau)) \right] \right),$$

(8.18)

where $H_{\mathrm{BH}}(b^*(\tau), b(\tau))$ is given by Eq. (8.16) with $\hat{b}_i \to b_i(\tau)$, and $\hat{b}_i^\dagger \to b_i^*(\tau)$. Solvability of the interacting model at $t_{ij} = 0$ suggests that instead of
the usual decoupling of the interaction term by the Hubbard–Stratonovich
transformation, this time we do the same with the hopping term. This leads
to the partition function expressed as an integral over the complex Hubbard–
Stratonovich field $\Phi_i(\tau)$:

$$Z = \int \prod_i \mathrm{D}\Phi_i^*(\tau)\mathrm{D}\Phi_i(\tau) e^{-S[\Phi]},$$

(8.19)

with the action

$$S[\Phi] = \int_0^\beta \mathrm{d}\tau \sum_{i,j} \Phi_i^*(\tau)(t^{-1})_{ij}\Phi_j(\tau)$$
$$- \sum_i \ln \int \mathrm{D}b^*(\tau)\mathrm{D}b(\tau) e^{-\int_0^\beta \mathrm{d}\tau L[b,\Phi_i]}.$$

(8.20)

We are assuming that the hopping matrix t_{ij} has all the eigenvalues positive.
The advantage of this strategy is that the Lagrangian,

$$L[b, \Phi_i] = -b^*(\tau)\partial_\tau b(\tau) - \mu|b(\tau)|^2 + \frac{U}{2}|b(\tau)|^2(|b(\tau)|^2 - 1)$$
$$- \Phi_i(\tau)b^*(\tau) - \Phi_i^*(\tau)b(\tau),$$

(8.21)

is completely local. This allows us to drop the site index on $b(\tau)$, since the
functional integral in the second term in Eq. (8.20) is the same at all sites.

We may next try to expand the action in powers of Φ and its imaginary time derivatives. Such an expansion will in principle have the form

$$S[\Phi] = S[0] + \int_0^\beta d\tau \left[\sum_{i,j} \Phi_i^*(\tau)(t^{-1})_{ij}\Phi_j(\tau) + \sum_i (u\Phi_i(\tau)^* \partial_\tau \Phi_i(\tau) \right.$$
$$\left. + v|\partial_\tau \Phi_i(\tau)|^2 + a|\Phi_i(\tau)|^2 + b|\Phi_i(\tau)|^4 + O(|\Phi_i|^6, \Phi^* \partial_\tau^3 \Phi)) \right].$$

$$(8.22)$$

The coefficients u, v, a, b are in principle functions of the chemical potential, the interaction, and the temperature. They may be obtained by computing the averages of the bosonic operators over the single-site Hamiltonian. To calculate the coefficients of the terms local in imaginary time, such as a and b, however, it suffices to take a time-independent Hubbard–Stratonovich field $\Phi_i(\tau) = \Phi_i$ in $L[b, \Phi]$. In this case Φ_i enters the single-site Lagrangian in Eq. (8.21) as an external static field. In the limit $T \to 0$ the action in Eq. (8.20) then becomes

$$S[\Phi] = \sum_{i,j} \Phi_i^*(t^{-1})_{ij}\Phi_j + \frac{1}{T}\sum_i E_0[\Phi_i], \qquad (8.23)$$

with $E_0[\Phi]$ as the ground state energy of the *single-site Hamiltonian*

$$\hat{H}[\Phi] = -\mu\hat{n} + \frac{U}{2}\hat{n}(\hat{n} - 1) - \Phi\hat{b}^\dagger - \Phi^*\hat{b}. \qquad (8.24)$$

This follows from the recognition of the functional integral that appears in Eq. (8.20) as the single-site partition function for the above Hamiltonian.

Elementary second-order perturbation theory for the ground state energy of $\hat{H}[\Phi]$ then gives

$$E_0[\Phi] = -\left(\mu + \frac{U}{2}\right)n + \frac{U}{2}n^2 + a|\Phi|^2 + b|\Phi|^4 + O(|\Phi|^6), \qquad (8.25)$$

with

$$a = \frac{n+1}{\mu - nU} + \frac{n}{-\mu + U(n-1)}. \qquad (8.26)$$

The relation between the chemical potential and the number of particles at each site given below Eq. (8.17) implies then that the coefficient a is always negative. One may similarly determine the coefficient b by going to the next order in the perturbation theory. For our present purposes it will be sufficient to know that the result for b is positive for all values of the chemical potential. Finite temperature corrections to the coefficients are then exponentially small

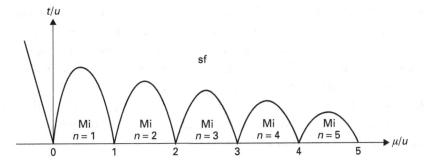

Figure 8.5 Phase diagram of the Bose–Hubbard model in the mean-field theory.

at low temperatures due to the finite gap in the spectrum of the single-site Hamiltonian $\hat{H}[\Phi]$.

For smooth configurations of $\Phi_i(\tau)$ we may take the limit of the vanishing lattice constant and finally write the continuum approximation to the action as

$$
\begin{aligned}
S[\Phi] = S[0] + \int_0^\beta d\tau \int d\vec{x} \big[& u\Phi^*(\vec{x}, \tau)\partial_\tau \Phi(\vec{x}, \tau) + v|\partial_\tau \Phi(\vec{x}, \tau)|^2 \\
& + w|\nabla\Phi(\vec{x}, \tau)|^2 + \tilde{a}|\Phi(\vec{x}, \tau)|^2 + b|\Phi(\vec{x}, \tau)|^4 \\
& + O(|\Phi|^6, \Phi^*\partial_\tau^3\Phi, |\nabla^2\Phi|^2)\big],
\end{aligned}
\tag{8.27}
$$

where we have also employed the usual gradient expansion. Defining $t(\vec{q})$ as the Fourier transform of the hopping matrix, the coefficient $\tilde{a}$ becomes

$$
\tilde{a} = \frac{1}{t} + a,
\tag{8.28}
$$

where $t = t(\vec{q} = 0)$. Neglecting the fluctuations, $\langle\Phi\rangle \neq 0$ for $\tilde{a} < 0$, or, when the hopping is larger than the critical value,

$$
\frac{t_c}{U} = \frac{(n - \frac{\mu}{U})(\frac{\mu}{U} - n + 1)}{\frac{\mu}{U} + 1}.
\tag{8.29}
$$

The reader can check that the average $\langle\Phi(x, \tau)\rangle \sim \langle b(x, \tau)\rangle$, so that the average value of the Hubbard–Stratonovich field is proportional to the superfluid order parameter. For $t > t_c$ the system is therefore in the superfluid phase, which is gapless and compressible. Below t_c the system is the Mott insulator with the gap. t_c therefore determines the mean-field superfluid–Mott insulator phase boundary at which the Mott gap collapses and superfluid order develops. It has the shape as shown in Fig. 8.5. Note that t_c vanishes at μ/U integer. Since at these special values of μ the ground state energy at $t = 0$ is macroscopically degenerate, an infinitesimal hopping is enough to select the superfluid ground state. Otherwise $t_c > 0$. There is therefore always a maximal t_c at a given n which is located near and below $\mu/U = n - (1/2)$.

So far we have confined the discussion to the determination of the coefficients of the terms local in imaginary time in the expansion in Eq. (8.22). This, however, turns out to be enough to fix all the remaining coefficients. Once the coefficient a has been found, for example, the coefficients u and v follow from the "gauge symmetry" of the action in Eq. (8.20). Notice that the global time-dependent transformation

$$b(\tau) \to b(\tau)e^{i\theta(\tau)}, \tag{8.30}$$

$$b^*(\tau) \to b^*(\tau)e^{-i\theta(\tau)}, \tag{8.31}$$

$$\Phi_i(\tau) \to \Phi_i(\tau)e^{i\theta(\tau)}, \tag{8.32}$$

$$\Phi_i^*(\tau) \to \Phi_i^*(\tau)e^{-i\theta(\tau)}, \tag{8.33}$$

$$\mu \to \mu + i\partial_\tau\theta(\tau), \tag{8.34}$$

leaves the form of the action invariant. Although the chemical potential under the gauge transformation becomes complex, this property may be exploited to find the coefficients u and v, for example. Demanding that the action in Eq. (8.27) is also gauge-symmetric after some algebra implies

$$\frac{\partial\tilde{a}}{\partial\mu} + u = 0, \tag{8.35}$$

$$\frac{\partial u}{\partial\mu} - 2v = 0, \tag{8.36}$$

and similarly for the coefficients of higher-order time derivatives. These equations are completely general. Using our particular result for $\tilde{a}$ yields

$$u = -\frac{1}{t_c^2}\frac{\partial t_c}{\partial\mu}, \tag{8.37}$$

and

$$v = \frac{1}{t_c^3}\left(\frac{\partial t_c}{\partial\mu}\right)^2 - \frac{1}{2t_c^2}\frac{\partial^2 t_c}{\partial\mu^2}, \tag{8.38}$$

with the critical hopping as the function of chemical potential as in Eq. (8.29). Note that the coefficient v is thus always positive, while u changes sign at the value of the chemical potential at which t_c is maximal. So whenever the superfluid transition may be tuned by changing the chemical potential, it is in the universality class of the $T = 0$ Bose–Einstein condensation. At the tip of the insulating lobe, on the other hand, the superfluid transition can be induced only by increasing hopping, and the mean-field theory suggests that it then lies in the universality class of the XY model in $d + 1$ dimensions.

The principal features of the phase diagram are in fact correct beyond the mean-field theory we used. First, since the Mott insulator has a gap in the spectrum at $t = 0$, it must take a finite hopping to collapse it, except at integer values of μ/U when the gap vanishes. Since the collapse of the gap coincides with the emergence of the superfluid order in the system, the exact superfluid–insulator phase boundary has the shape qualitatively similar to the mean-field result. Second, changing hopping inside the insulating phase cannot change the density of particles. Since the hopping part of the Hamiltonian commutes with total particle number, the ground state at small $t \neq 0$ is still an eigenstate of the total number of particles. Since there is a finite gap, on the other hand, the total number of particles in the ground state at small $t \neq 0$ and $t = 0$ can differ only by an amount proportional to t/U. Discreteness of the particle number then implies that it remains constant as long there is a gap in the spectrum. Since the particle density within a lobe thus remains the same along the line $t = 0$, and cannot change by increasing hopping as long as there is a gap in the spectrum, the whole insulating lobe corresponds to the same integer particle number per site.

The nature of the superfluid–insulator quantum phase transition may be inferred as follows. Since the gauge invariance is an exact symmetry in the theory, the general Eqs. (8.35) and (8.36) must continue to hold between the renormalized coefficients in the low-energy action, which would be obtained by integrating out the high-energy fluctuations of the order parameter. The transition is then still determined by the condition $\tilde{a}(\mu, t) = 0$, but with the exact $\tilde{a}(\mu, t)$ being fully renormalized by the fluctuations. It is easy to see that the coefficient u then still has to vanish at the tip of the insulating lobe. If we consider the two-dimensional gradient of the function $\tilde{a}(\mu, t)$, at the tip of any line of constant $\tilde{a}(\mu, t)$ the gradient will become orthogonal to the μ-axis, so that $\partial \tilde{a}(\mu, t)/\partial \mu = 0$ at the line of such tips. Gauge invariance of the action implies then that at such a line of tips the coefficient $u = 0$. So there exists a line in the μ–t plane that goes through the tip of the insulating lobe along which the superfluid–insulator transition is indeed in the XY universality class. In the mean field theory this line is approximated by a straight line parallel to the hopping axis. Away from this line the transition is in the universality class of the $T = 0$ Bose–Einstein condensation.

It is useful to consider the line of constant integer density in the superfluid phase. First, this line also must meet the insulating lobe precisely at the tip: assuming that it does not would lead to a negative compressibility in a finite region near the tip, which is physically impossible. Second, the density of bosons as a function of the chemical potential and the hopping, $N(\mu, t)$, being a fixed integer everywhere inside the insulating lobe, is such that the sign of $\partial N(\mu, t)/\partial t$ at $t = t_c + 0$ changes from negative to positive at the phase boundary as one goes around the lobe in the direction of increasing chemical potential. It therefore has to vanish somewhere at the phase boundary. At that point increasing hopping by an infinitesimal amount does not change the integer density of particles. Since there could be only one such line with the integer density ending at the given insulating lobe, and that right at the tip, at the tip $\partial N(\mu, t)/\partial t = 0$. The line of integer density meets the tip orthogonally to the axis of the chemical potential. One can show explicitly that this indeed is what happens within the mean-field theory. The line of integer density and the line $u = 0$, however, are in principle different.

Problem 8.3 Show that the Bose–Hubbard model has the superfluid ground state at integer μ/U for any value of hopping by using the degenerate perturbation theory.

Solution When $\mu/U = n$ and $\Phi = 0$, the single-site Hamiltonian $\hat{H}[0]$ in Eq. (8.24) has two degenerate ground states, $|n\rangle$ and $|n + 1\rangle$. In this basis $\hat{H}[\Phi]$ is represented by a two by two matrix with $-nU(n + 1)/2$ on the diagonal, and $-\sqrt{n + 1}\Phi$ and its complex conjugate as off-diagonal elements. The ground state energy is therefore

$$E_0[\Phi] = -\frac{U}{2}n(n + 1) - (n + 1)|\Phi| + O(|\Phi|^2).$$

Since the positive hopping term in $S[\Phi]$ in Eq. (8.23) is quadratic in Φ, the value of the action can always be decreased by developing a weak superfluid order.

Problem 8.4* Compute the coefficients u and v in Eq. (8.27) without the use of the gauge symmetry.

Solution Expanding the exponential in Eq. (8.20) in powers of Φ to quadratic order yields

$$S[\Phi(\tau)] = S[0] + \int_0^\beta d\tau \sum_{i,j} \Phi_i^*(\tau)(t^{-1})_{ij}\Phi_j(\tau)$$
$$- \sum_i \int \frac{d\omega}{2\pi} \langle b^*(\omega)b(\omega)\rangle |\Phi_i(\omega)|^2,$$

where the average is supposed to be taken over the Lagrangian in Eq. (8.21) with $\Phi_i = 0$. The average is

$$\langle b^*(\omega)b(\omega)\rangle = \int_{-\infty}^{\infty} d(\tau_2 - \tau_1)e^{-i\omega(\tau_2 - \tau_1)} \langle n|T_\tau\{\hat{b}(\tau_1), \hat{b}^\dagger(\tau_2)\}|n\rangle,$$

where $|n\rangle$ is the ground state of the single-site Hamiltonian $\hat{H}[0]$ in Eq. (8.24), and $\hat{b}(\tau) = e^{\hat{H}[0]\tau}\hat{b}e^{-\hat{H}[0]\tau}$, $\hat{b}^\dagger(\tau) = e^{\hat{H}[0]\tau}\hat{b}^\dagger e^{-\hat{H}[0]\tau}$. T_τ is the time-ordering operator that places operators at earlier times to the right of those at later times. So,

$$\langle b^*(\omega)b(\omega)\rangle = \int_0^{\infty} d\tau e^{-i\omega\tau} \langle n|\hat{b}^\dagger(\tau)\hat{b}(0)|n\rangle + \int_0^{\infty} d\tau e^{i\omega\tau} \langle n|\hat{b}(\tau)\hat{b}^\dagger(0)|n\rangle$$

$$= \frac{n}{i\omega + \mu - U(n-1)} + \frac{n+1}{-i\omega - \mu + nU}.$$

At $\omega = 0$ this reproduces the mean-field expression for the coefficient a in Eq. (8.26). The Taylor expansion in powers of frequency yields the coefficients u and v, in accordance with the gauge symmetry.

8.4 Quantum fluctuations and the superfluid density

It is sometimes convenient to express the complex functions $b_i(\tau)$, $b_i^*(\tau)$ in the partition function of the Bose–Hubbard model in Eq. (8.18) in terms of their amplitude and the phase:

$$b_i(\tau) = \rho_i^{1/2}(\tau)e^{i\phi_i(\tau)}. \tag{8.39}$$

In the superfluid phase the variable $\rho_i(\tau)$ can be viewed as the fluctuating local density of particles. Writing $\rho_i(\tau) = \rho_0 + \delta\rho_i(\tau)$, with $\rho_0 = \mu/U + 1/2$ as the average particle number per site, and expanding the action in Eq. (8.18), one finds $S = \int_0^\beta d\tau L$ with the Lagrangian

$$L = i(\rho_0 + \delta\rho_i(\tau)) \sum_i \dot{\phi}_i(\tau)$$

$$- \rho_0 \sum_{\langle ij \rangle} t_{ij} \cos(\phi_i(\tau) - \phi_j(\tau)) + \frac{U}{2} \sum_i \delta\rho_i^2(\tau). \tag{8.40}$$

Here we assumed nearest-neighbor hopping and neglected the imaginary time derivative of the density, as well as deviations from the average density in the hopping term. The latter approximation is justified if $\rho_0 \gg 1$.

The imaginary time in the Lagrangian that couples local density to the time derivative of the local phase is reminiscent of the term in the Feynman's path integral for a quantum mechanical particle, with the density playing the role of the momentum, and the phase of the position observable. By analogy, then, the density and the phase may be considered as canonically conjugate variables, satisfying an analogous commutation relation. This provides an alternative view at the superfluid–insulator transition from the last section. In the insulating phase the quantum uncertainty of the local density of particles is small. The uncertainty relation implies then that the fluctuation of the canonically conjugate phase variable must be large. Similarly to thermal fluctuations at finite temperature, these *quantum fluctuations* of the phase may be enough to destroy the long-range order in the system. In the superfluid phase, on the other hand, the situation becomes reversed.

Quantum fluctuations become most apparent if we integrate over the small density deviations $\delta\rho_i$ in L. In terms of the remaining phase variable the Lagrangian then reduces to

$$L = i\rho_0 \sum_i \dot{\phi}_i(\tau) - \rho_0 \sum_{\langle ij \rangle} t_{ij} \cos(\phi_i(\tau) - \phi_j(\tau)) + \frac{1}{2U} \sum_i \dot{\phi}_i{}^2(\tau). \quad (8.41)$$

When $U = 0$ the fluctuations of $\phi_i(\tau)$ in imaginary time become completely suppressed, and the partition function in the case of nearest-neighbor hopping reduced to the XY model in Eq. (6.29), with $J = t\rho_0$. When U is large, on the other hand, imaginary time fluctuations of the phase become significant. Furthermore, once the fluctuations in imaginary time become allowed the Lagrangian is no longer always real. Since the periodic boundary conditions for $b_i(\tau)$ imply that $\phi_i(\beta) = \phi_i(0) + 2\pi n_i$, the action is real only at an integer density. Note that, for the assumed large ρ_0, this happens at the tip of the insulating lobe, located at $\mu/U = n - 1/2$ for large n.

One may further understand the role of quantum fluctuations by computing the superfluid density, i.e. the stiffness of the phase, using the variational method in which the action defined by the above Lagrangian is replaced by the optimal quadratic one (Problem 6.4):

$$S_0 = \frac{T}{2} \sum_{\omega_n, \vec{k}} \phi(\omega_n, \vec{k}) G_0^{-1}(\omega_n, \vec{k}) \phi(-\omega_n, -\vec{k}). \quad (8.42)$$

S_0 is to be chosen to minimize the variational free energy $F_{\text{var}} = F_0 + T \langle S - S_0 \rangle_0$, with the last average taken over the ensemble defined by S_0. At low temperatures only small fluctuations in imaginary time are important, so that all the finite winding numbers $n_i \neq 0$ may be neglected. We may therefore drop the imaginary term in the action. The variational free energy is then easily computed:

$$F_{\text{var}} = -\frac{T}{2} \sum_{\omega_n, \vec{k}} \ln G_0(\omega_n, \vec{k}) + \frac{T}{2U} \sum_{\omega_n, \vec{k}} \omega_n^2 G_0(\omega_n, \vec{k}) \tag{8.43}$$

$$-J \sum_{\hat{r}} e^{-T \sum_{\omega_n, \vec{k}} G_0(\omega_n, \vec{k})(1 - \cos(\vec{k} \cdot \hat{r}))},$$

where $\hat{r}$ is a lattice unit vector, and we assumed the nearest-neighbor hopping matrix t_{ij}. Minimizing with respect to $G_0(\omega_n, \vec{k})$ then yields

$$G_0^{-1}(\omega_n, \vec{k}) = \frac{\omega_n^2}{U} + 4K \sum_{\hat{r}} \sin^2\left(\frac{\vec{k} \cdot \hat{r}}{2}\right), \tag{8.44}$$

with K as

$$K = J \exp\left(-\frac{1}{2d\sqrt{K}} \int \frac{d\vec{q}}{(2\pi)^2} f(\vec{q}) \coth\left(\frac{\sqrt{K}}{T} f(\vec{q})\right)\right), \tag{8.45}$$

with

$$f^2(\vec{k}) = U \sum_{\hat{r}} \sin^2\left(\frac{\vec{k} \cdot \hat{r}}{2}\right). \tag{8.46}$$

The temperature-dependent quantity K evidently is the coefficient of the $(\nabla\phi)^2$ term in S_0, and should therefore be identified with the phase stiffness. At $T = 0$ the superfluid density is reduced by the quantum fluctuations as

$$\frac{K(0)}{J} = 1 - \frac{1}{2dJ} \int \frac{d\vec{q}}{(2\pi)^2} f(\vec{q}) + O(U). \tag{8.47}$$

At $U = 0$, on the other hand, K in the XY model at low temperatures satisfies

$$K = J e^{-\frac{T}{2dK}} \approx J - \frac{T}{2d} + O(T^2). \tag{8.48}$$

The last result agrees well with Monte Carlo calculations, including the coefficient of the term linear in temperature. Finally, for finite U and at temperatures well below the temperature scale for quantum fluctuations $\omega_q^{2(d+1)} = U^d K(0)^{d+2}$,

$$K \approx K(0)\left(1 - I(d)\left(\frac{T}{\omega_q}\right)^{d+1}\right), \tag{8.49}$$

with the coefficient $I(d)$ given by the integral

$$I(d) = \frac{1}{2d} \int \frac{d\vec{q}}{(2\pi)^d} q e^{-q}. \tag{8.50}$$

The interactions thus have a twofold effect. First, by introducing phase fluctuations in the direction of the imaginary time they reduce the stiffness of the phase even at $T = 0$. Second, they suppress further reduction of the stiffness with temperature by introducing the quantum energy scale ω_q below which the power in the power-law behavior of phase stiffness vs. temperature becomes higher than one. The $\sim T^4$ dependence in three dimensions is in excellent agreement with the measurements of the superfluid density in ^{4}He.

Problem 8.5 Estimate the critical temperature of the XY model by using the variational result for the superfluid density.

Solution Using the universal value of the superfluid density jump at the Kosterlitz–Thouless transition, $K(T_c)/T_c = 2/\pi$ in Eq. (6.80) gives

$$\frac{T_c}{J} = \frac{\pi}{2} e^{-\frac{\pi}{8}} = 1.06.$$

Monte Carlo simulations give an essentially exact result, $T_c/J = 0.90$, for the two-dimensional XY model. The variational approximation overestimates the critical temperature primarily due to its neglect of vortices, which are very efficient in reducing the phase stiffness at temperatures near T_c.

Problem 8.6 Find the reduction of the superfluid density at low temperatures in two and three dimensions in the case of Coulomb interaction.

Solution Generalizing to an arbitrary interaction, the variational approximation gives F_{var} as in Eq. (8.43) with $1/U$ replaced with $1/V(\vec{k})$ under the momentum sum. This yields the stiffness K being determined by the same Eq. (8.45), except that

$$f^2(\vec{k}) = V(k) \sum_{\hat{r}} \sin^2 \left(\frac{\vec{k} \cdot \hat{r}}{2} \right).$$

In two dimensions $V(k) = e^2/|k|$ at small momenta, and

$$\frac{K(T)}{K(0)} = 1 - \frac{6}{\pi} \left(\frac{T}{\omega_q} \right)^5.$$

with $\omega_q = e^{4/5} K^{3/5}(0)$. Similarly, in three dimensions $V(k) = e^2/k^2$ so that

$$K(0) - K(T) \propto \omega_q e^{-\frac{\omega_q}{T}},$$

with $\omega_q^2 = e^2 K_0$ being the plasma frequency.

8.5 Universal conductivity in two dimensions*

If we consider the bosons in the Bose–Hubbard model as possessing an electrical charge e^*, further interesting results for the conductivity of the system in two dimensions may be deduced. If we are interested only in the conductivity near the superfluid–insulator transition, we may compute it as a response of the system to a weak external vector potential $\vec{a}(\vec{r}, \tau)$, with $\vec{a}(\vec{r}, \tau)$ entering the partition function via the usual replacement

$$\nabla \rightarrow \nabla - i\frac{e^*}{\hbar}\vec{a}(\vec{r}, \tau) \tag{8.51}$$

in the action for the superfluid order parameter in Eq. (8.27). The equality of the charge of the Hubbard–Stratonovich field Φ and the original bosons follows from the gauge invariance of the action in Eq. (8.20). The average electrical current is then

$$\langle j_i(\vec{r}, \tau)\rangle = -\hbar\frac{\delta F}{\delta a_i(\vec{r}, \tau)}, \tag{8.52}$$

where $F = -\ln Z$, and Z is the partition function. The conductivity tensor σ_{ij} may then be defined at imaginary times as determining the response to an external electric field,

$$\langle j_i(\vec{r}, \tau)\rangle = \int d\vec{r}'d\tau'\sigma_{ij}(\vec{r} - \vec{r}', \tau - \tau')E_j(\vec{r}', \tau'), \tag{8.53}$$

where the electric field is

$$E_i(\vec{r}, \tau) = -i\frac{\partial a_i(\vec{r}, \tau)}{\partial \tau}. \tag{8.54}$$

The last equation becomes the equivalent of the standard relation between the electric field and the electromagnetic vector potential in real time t upon substitution $\tau = it$.

Performing a partial integration in Eq. (8.53) and then differentiating with respect to the vector potential once again, we find

$$\frac{1}{i}\frac{\partial\sigma_{ij}(\vec{r} - \vec{r}', \tau - \tau')}{\partial\tau'} = \hbar\frac{\delta^2 F}{\delta a_j(\vec{r}', \tau')\delta a_i(\vec{r}, \tau)}. \tag{8.55}$$

Evaluation of the derivative on the right hand side at $\vec{a}(\vec{r}, \tau) \equiv 0$ gives then the linear conductivity. Its uniform ($\vec{q} = 0$) component is then

$$\sigma_{ij}(q = 0, \omega) = \frac{\rho_{ij}(\omega)}{\omega}\frac{e^{*2}}{\hbar}, \tag{8.56}$$

with the function $\rho_{ij}(\omega)$ as

$$\rho_{ij}(\omega) = 2w\delta_{ij}\langle\Phi^*(0,0)\Phi(0,0)\rangle - w^2 \int d\vec{r} \int_0^\beta d\tau e^{i\omega\tau}\langle j_i(\vec{r},\tau)j_j(0,0)\rangle$$

(8.57)

and the current $\vec{j}(\vec{r},\tau) = i\Phi^*(\vec{r},\tau)\nabla\Phi(\vec{r},\tau) - i\Phi(\vec{r},\tau)\nabla\Phi^*(\vec{r},\tau)$. The last result is an example of the general Kubo formula for the linear response to an external perturbation. The two terms appearing in $\rho_{ij}(\omega)$ are often called the "diamagnetic" and the "paramagnetic" contributions, respectively. The response in real physical time may be obtained from the knowledge of $\sigma_{ij}(\omega)$ by performing the analytical continuation from Matsubara to real frequencies in the last equation, as $i\omega \rightarrow \omega + i\eta$.

The form of the conductivity near the transition may be easily derived by noticing that Eq. (8.57) implies that it scales with length as $\sigma \sim L^{d+2z}j^2$. Scaling of the electrical current j, on the other hand, may be determined exactly, since the gauge invariance dictates that the form $\nabla - (e/\hbar)\vec{a}$ remains preserved during the momentum shell transformation, so that $\vec{a} \sim L^{-1}$. Since $jaL^{d+z} \sim 1$, we find that $\sigma \sim L^{2-d}$. In an isotropic system $\sigma_{ij}(q = 0, \omega) = \sigma(\omega)\delta_{ij}$, and we can write

$$\sigma(\omega) = \xi^{2-d} F_\pm\left(\frac{\hbar\omega}{k_B T}, \frac{\hbar\omega_0|\delta|^{z\nu}}{k_B T}\right)\frac{(e^*)^2}{\hbar},$$

(8.58)

where $\xi \sim 1/|\delta|^\nu$ is the correlation length, $\hbar\omega_0$ is the characteristic microscopic energy scale, and δ the dimensionless tuning parameter for the superfluid–insulator transition. $F_\pm$ are two universal dimensionless functions of two dimensionless variables, corresponding to the two sides of the transition.

Of course, the scaling form for the conductivity is just a special case of the general scaling relations near critical points. What is remarkable about it, however, is that we were able to determine the dimensionality of conductivity, namely the power $2 - d$, exactly. The anomalous dimension of the current vanishes, and its scaling with length is determined exactly by its engineering dimension. This is a consequence of the gauge-invariant way the vector potential enters the action. Another example of the same phenomenon was provided by the exact result for the anomalous dimension of the gauge field in the Ginzburg–Landau theory in Section 4.4.

The scaling relation in Eq. (8.58) has an interesting and a rather non-intuitive physical consequence. In $d = 2$ the exponent of the correlation length dependent prefactor vanishes, and at the critical point at $\delta = 0$,

$$\sigma_c(\omega) = f\left(\frac{\hbar\omega}{k_B T}\right)\frac{(e^*)^2}{\hbar}, \tag{8.59}$$

where $f(x) = F_\pm(x, 0)$ is a universal function. In particular, the dc conductivity at the superfluid transition is a universal number in units of $(e^*)^2/\hbar$! If this number is finite, this implies that at $T = 0$, and right at the critical point between insulating and superfluid phases the system becomes a curious metal, the conductivity of which is a constant of nature. Something resembling this phenomenon is observed in thin metallic films with thickness used as a tuning parameter. Assuming that near the quantum critical point Cooper pairs are the effective bosons in the problem, the charge e^* is twice the electron charge, and $\hbar/e^{*2} \approx 1000$ ohm. The experiment suggest then that $f(0) \approx 0.8 - 0.9$. This number, however, is not directly relevant to the Bose–Hubbard model, since the long-range Coulomb interaction and disorder, both present in real experiments, change the universality class of the superfluid–insulator transition.

The difficulties of the computation of the function $f(x)$ may be appreciated by considering the two opposite limits of $\hbar\omega/k_B T \ll 1$ and $k_B T/\hbar\omega \ll 1$, and by realizing that they are determined by two entirely different mechanisms of dissipation. The former corresponds to the transport of pre-existing thermally excited quasiparticles, which get accelerated by the electric field, and relax back to equilibrium via mutual collisions. The system is therefore in the hydrodynamic regime, and the transport is in principle described by the appropriate Boltzmann equation, similarly to the Drude conductivity of a metal. It is indeed quite remarkable that the conductivity in this limit of completely incoherent transport is nevertheless universal. The latter limit, on the other hand, corresponds to the collisionless transport of coherent excitations created by the electric field. The two limits of $f(x)$ therefore may be expected to be different, and indeed calculations show that $f(0) > f(\infty)$. It is important to realize that both limits in principle describe the transport in the system at low temperatures. The experiments, however, typically lie in the hydrodynamic regime where $\hbar\omega/k_B T \ll 1$, and the measurements should in principle yield the value of $f(0)$.

Computation of $f(x)$ may be performed using the $\epsilon = 3 - d$ expansion, or the large-N expansion in $d = 2$. It becomes, however, quite technical, and we will describe here only the evaluation of the leading term in ϵ-expansion, which will be sufficient to get the correct physical picture. First, assume that the electromagnetic potential, and therefore the current, is along the x-direction. Neglecting the interaction term in the action in Eq. (8.27), at finite temperatures we can write

$$\rho(\omega) = 2c^2 k_B T \sum_{\nu_n} \int \frac{d\vec{k}}{(2\pi)^d} \left(\frac{1}{\nu_n^2 + c^2 k^2 + m^2} \right.$$

$$\left. - \frac{2c^2 k_x^2}{\left(\nu_n^2 + c^2 k^2 + m^2\right)((\nu_n + \omega)^2 + c^2 k^2 + m^2)} \right), \quad (8.60)$$

where $c^2 = w/v$, $m^2 = \tilde{a}/v$, and we assumed $u = 0$ for convenience. The last equation therefore gives the conductivity at the superfluid–insulator transition in the XY universality class, to the leading order in small $\epsilon = 3 - d$. The expression may be simplified by inserting the unity under the first integral written as $\partial k_x / \partial k_x = 1$, and then integrating by parts. Since the boundary term vanishes, we may write

$$\rho(\omega) = \sum_{\nu_n} \int \frac{d\vec{k}}{(2\pi)^d} \frac{4c^4 k_B T k_x^2}{\nu_n^2 + c^2 k^2 + m^2}$$

$$\times \left(\frac{1}{\nu_n^2 + c^2 k^2 + m^2} - \frac{1}{(\nu_n + \omega)^2 + c^2 k^2 + m^2} \right). \quad (8.61)$$

Written in this form, it becomes evident that $\rho(\omega)$ vanishes at small Matsubara frequency. Performing the sums over the Matsubara frequencies, we find $\sigma(\omega) = \sigma_1(\omega) + \sigma_2(\omega)$, with

$$\sigma_1(\omega) = \frac{2c^{2-d}}{\omega} \int \frac{d\vec{k}}{(2\pi)^d} \frac{k_x^2}{e_k^2} \left(-\frac{\partial n_B(e_k)}{\partial e_k} \right) \frac{e^{*2}}{\hbar}, \quad (8.62)$$

$$\sigma_2(\omega) = \omega c^{2-d} \int \frac{d\vec{k}}{(2\pi)^d} \frac{k_x^2}{e_k^3(\omega^2 + 4e_k^2)} (1 + 2n_B(e_k)) \frac{e^{*2}}{\hbar}, \quad (8.63)$$

and $e_k^2 = k^2 + m^2$.

The two contributions to conductivity have very different physical origins. Analytically continuing to real frequencies and using the identity

$$\lim_{\eta \to 0} \frac{1}{\omega + i\eta} = P \left(\frac{1}{\omega} \right) - i\pi \delta(\omega), \quad (8.64)$$

in $d = 3$ we obtain

$$\text{Re}\sigma_1(\omega) = \frac{(k_B T)^2}{3\pi c} \delta(\omega) f_1 \left(\frac{m}{k_B T} \right) \frac{(e^*)^2}{\hbar}, \quad (8.65)$$

with

$$f_1(z) = \int_0^\infty dt \frac{t^4}{t^2 + z^2} \frac{e^{\sqrt{z^2 + t^2}}}{(e^{\sqrt{z^2 + t^2}} - 1)^2} = \frac{\pi^2}{3} + O(z). \quad (8.66)$$

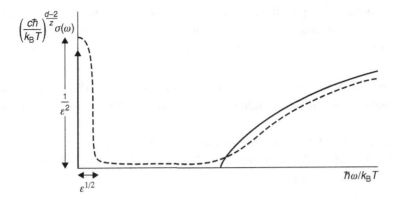

Figure 8.6 Conductivity in the insulating phase: the full line represents the result in the absence of interactions, with the heavy arrow at $\omega = 0$ as the delta-function. The dashed line is the result to second order in $\epsilon = 3 - d$.

For $m \ll k_B T$ and in $d = 3$, therefore,

$$\text{Re}\sigma_1(\omega) = \frac{\pi}{9} \left(\frac{k_B T}{c\hbar} \right) \delta \left(\frac{\hbar\omega}{k_B T} \right) \frac{e^{*2}}{\hbar}. \tag{8.67}$$

Similarly, analytically continuing to real frequencies one finds

$$\text{Re}\sigma_2(\omega) = \frac{1}{48\pi} \frac{\omega}{c} \left(\frac{\omega^2 - 4m^2}{\omega^2} \right)^{3/2} \left(1 + 2n_B \left(\frac{\hbar\omega}{2} \right) \right) \Theta(\omega - 2m). \tag{8.68}$$

Both terms indeed acquire the announced scaling form upon identification of the correlation length as $\xi = 1/m$, with the dynamical exponent $z = 1$. σ_2 vanishes below the threshold for the creation of a particle–hole pair, and describes the collisionless transport of coherent field-induced quasiparticles. σ_1, on the other hand, albeit vanishing at zero temperature, becomes proportional to $\delta(\omega)$ at finite temperatures. This part in principle describes the collision-dominated transport of incoherent, thermally generated excitations. In the Gaussian approximation, however, the excitations are completely coherent, and yield therefore the response of an ideal conductor. Including interactions will provide them with finite lifetimes, and one expects that the delta-function at $\omega = 0$ will be spread into a Lorentzian, as in Fig. 8.6. This is indeed what is found to the second order in interactions. The universality of the collision-dominated transport derives from the fact that the interaction itself becomes a universal number of order ϵ near the quantum critical point.

Problem 8.7* Compute the universal conductivity at the superfluid–insulator transition in two dimensions, in the regime $\hbar\omega/k_B T \gg 1$, in Gaussian approximation.

Solution In the collisionless limit we may set $T = 0$ in Eq. (8.63). At the critical point $m = 0$, and the straightforward integration gives $\sigma_2 = (1/16)(e^{*2}/\hbar)$.

Problem 8.8* Determine the lowest order correction in $\epsilon = 3 - d$ to the weight of the delta-function peak in $\sigma_1(\omega)$, at a temperature T right above the quantum critical point.

Solution First, let us determine the shift in the critical point at finite temperature. To the lowest order in interaction, we find

$$m^2(T) = -\mu + 4\lambda T \sum_{\omega_n} \frac{1}{\omega_n^2 + k^2 - \mu} + O(\lambda^2),$$

where we set $u = 0$, $v = 1$ in the action in Eq. (8.8). Adding and subtracting the $T = 0$ contribution, we may write

$$m^2(T) = m^2(0) + 4\lambda \int \frac{d\vec{k}}{(2\pi)^d} \frac{n_B(\sqrt{k^2 + m^2(0)})}{\sqrt{k^2 + m^2(0)}} + O(\lambda^2).$$

At the quantum critical point at $m(0) = 0$, therefore,

$$m^2(T) = \frac{4\lambda S_d}{(2\pi)^d} T^{d-1} \int_0^\infty \frac{k^{d-2}dk}{e^k - 1}.$$

To the lowest order in $\epsilon = 3 - d$ we need to evaluate the above integral at $d = 3$, and replace $\hat{\lambda}$ as defined below Eq. (8.11) with its fixed point value, $\hat{\lambda} = \epsilon/10$. At the critical point, therefore,

$$\frac{m(T)}{T} = \frac{2\pi}{\sqrt{15}} \epsilon^{1/2} + O(\epsilon^{3/2}).$$

Expanding, on the other hand, the function $f_1(z)$ in Eq. (8.66) around $z = 0$ yields

$$f_1(z) = \frac{\pi^2}{3}\left(1 - \frac{9}{4\pi}z + O(z^2)\right),$$

so that

$$\mathrm{Re}\sigma_1(\omega) = \frac{\pi}{9}\left(1 - \frac{9}{2\sqrt{15}}\sqrt{\epsilon} + O(\epsilon)\right)\left(\frac{k_B T}{c\hbar}\right)\delta\left(\frac{\hbar\omega}{k_B T}\right)\frac{e^{*2}}{\hbar}.$$

Problem 8.9* Determine the relevance of Coulomb interaction between bosons at the quantum transition to a superfluid state in two dimensions.

Solution Let us add the term

$$\int_0^{1/T} d\tau \int d\vec{x}\, d\vec{y} |\Phi(\vec{x}, \tau)|^2 \frac{e^2}{2|\vec{x} - \vec{y}|} |\Phi(\vec{y}, \tau)|^2$$

to the action for interacting bosons in Eq. (8.8), with $u = 1$ and $v = 0$. Introducing the Hubbard–Stratonovich scalar gauge field $a(\vec{x}, \tau)$ to decouple the Coulomb interaction, we may write the action at $T = 0$ as

$$S = \int d\tau d\vec{x} [\Phi^*(\vec{x}, \tau)(\partial_\tau - ia(\vec{x}, \tau))\Phi(\vec{x}, \tau) + |\nabla\Phi(\vec{x}, \tau)|^2$$
$$-\mu|\Phi(\vec{x}, \tau)|^2 + \lambda|\Phi(\vec{x}, \tau)|^4 + \frac{1}{2e^2}a(\vec{x}, \tau)|\nabla|a(\vec{x}, \tau),$$

with the operator $|\nabla|$ to be understood as $|\vec{q}|$ in Fourier space. The neutralizing background can be included by omitting the $\vec{k} = 0$ mode for the gauge field. In this form the first two terms in the action are invariant under a limited gauge transformation

$$\Phi(\vec{x}, \tau) \to e^{i\phi(\tau)}\Phi(\vec{x}, \tau),$$
$$a(\vec{x}, \tau) \to a(\vec{x}, \tau) + \partial_\tau \phi(\tau).$$

This guarantees that their form remains invariant under the momentum-shell transformation. So the gauge field scales exactly as $a \sim L^{-z}$ with length, which then implies $e^2 \sim L^{1-z}$.

The correction to the order parameter quadratic action still vanishes. Also, since the quadratic term for the gauge field is non-analytic in momentum, it cannot renormalize. This implies that the β-functions for the two interaction couplings are

$$\frac{d\hat{e}^2}{d\ln b} = (z - 1)\hat{e}^2,$$

with $\hat{e}^2 = e^2\Lambda^{1-z}$ and with $z = 2$, and Eq. (8.14) with $\epsilon = 0$. Whereas the short-range interaction is marginally irrelevant in two dimensions, a weak Coulomb interaction is relevant. This should be interpreted as that the system suffers a first-order transition from the superfluid state into the Wigner crystal as the density of bosons is decreased.

Problem 8.10* Determine the universality class of the superfluid transition in the system of lattice bosons at a commensurate density and in general

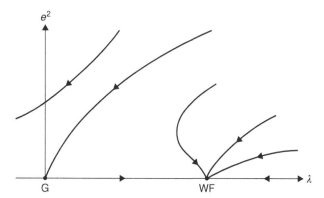

Figure 8.7 Flow on the critical surface for bosons interacting via Coulomb interaction at a commensurate density in dimension $d < 3$. The transition may be continuous or discontinuous depending on the value of the charge.

dimension $d \leq 3$, if the bosons are interacting both via short-range and Coulomb interactions.

Solution At a commensurate density the limited gauge-invariance observed in the previous problem dictates that the action in the presence of a Coulomb interaction between bosons may be written as

$$S = \int d\tau d\vec{x} [|(\partial_\tau - ia(\vec{x}, \tau))\Phi(\vec{x}, \tau)|^2 + |\nabla\Phi(\vec{x}, \tau)|^2$$
$$- \mu|\Phi(\vec{x}, \tau)|^2 + \lambda|\Phi(\vec{x}, \tau)|^4 + \frac{1}{2e^2} a(\vec{x}, \tau)|\nabla|^{d-1} a(\vec{x}, \tau),$$

in d dimensions. Integrating the fast modes with $\Lambda/b < k < \Lambda$ and $-\infty < \omega < \infty$ leads to changing the coefficients in front of ω (Z_ω), k^2 (Z_k), μ (Z_μ), λ (Z_λ), and a-term (Z_a) in the Fourier transformed action. One then rescales the momentum as $bk \to k$, the frequency as $b^z\omega \to \omega$, and the fields as $b^{-d}a \to a$ and $b^{-(2+d+z)/2} Z_k^{1/2}\Phi \to \Phi$ to find finally that the action at the critical point can be restored into its original form provided that the coupling constants are changed into $\lambda(b) = b^{4-d-z} Z_k^{-2} Z_\lambda \lambda$ and $e^2(b) = b^{z-1} Z_a^{-1} e^2$. Computing the Z-factors to the leading order then yields

$$\beta_\lambda = \left(\epsilon + \frac{e^2}{2}\right)\lambda - \frac{5}{2}\lambda^2 - \frac{1}{4}e^4,$$

$$\beta_e = (z - 1)e^2 - \frac{\delta_{3,d}}{12}e^4,$$

where $\epsilon = 3 - d$. The dynamical exponent is fixed by requiring that $b^{-2z} Z_\omega = b^{-2} Z_k$, so that

$$z = 1 - \frac{1}{3}e^2.$$

The last term in β_e is finite only when $d = 3$, and the quadratic term for the gauge field becomes analytic in momentum.

Small charge is therefore irrelevant at the Wilson–Fisher critical point at $e^2 = 0$ and $\lambda = 2\epsilon/5$, when $\epsilon > 0$. The transition for weak charge therefore remains continuous, in the XY universality class, and with $z = 1$. For large charge, on the other hand, the transition becomes fluctuation-induced first order. This follows from observing that the flow originating on the line $\lambda = 0$ is towards negative λ, so there must exist a separatrix that divides the flow diagram into the regions of first and second order transitions. This situation is depicted in Fig. 8.7.

Finally, precisely in $d = 3$ one can find the flow trajectories explicitly:

$$C + \ln e^2 = \frac{60}{239} \arctan\left(\frac{60(\lambda/e^2) - 11}{\sqrt{239}}\right),$$

where C is a constant. All solutions run away to the region with negative λ, and the transition is consequently first order for all values of the parameters.

Appendix A

Hubbard–Stratonovich transformation

Consider the Gaussian integral over a single real variable x:

$$\int_{-\infty}^{\infty} dx e^{-\frac{a}{2}x^2 + xz} = \left(\frac{\pi}{a}\right)^{1/2} e^{\frac{z^2}{2a}},$$

with $a > 0$. This can be readily generalized into

$$\int_{-\infty}^{\infty} \left(\prod_{i=1}^{N} dx_i\right) e^{-\frac{1}{2}\sum_{i,j=1}^{N} x_i A_{ij} x_j + \sum_{i=1}^{N} x_i z_i} = \left(\frac{\pi}{\det A}\right)^{1/2} e^{\frac{1}{2}\sum_{i,j=1}^{N} z_i A_{ij}^{-1} z_j},$$

where A^{-1} is the inverse of the real matrix A, assumed to have all the eigenvalues real and positive. The last equation is easily proved by rotating into the eigenbasis of the matrix A, and then by using the previous identity for Gaussian integrals.

Generalizing further to functional integrals, we can write

$$\int D\chi(x) e^{-\frac{1}{2}\int dxdy \chi(x) K^{-1}(x-y)\chi(y) + \int dx \chi(x)\Psi(x)} = e^{\frac{1}{2}\int dxdy \Psi(x) K(x-y)\Psi(y)},$$

where x is a multidimensional coordinate. We omitted the constant of proportionality, and assumed it is being absorbed into the measure of the functional integral. The kernel $K(x - y)$ is again assumed to have positive eigenvalues. The last identity follows from writing the integral in the eigenbasis of the kernel $K(x - y)$ and then by applying the previous identity for multidimensional Gaussian integrals.

The form of Hubbard–Stratonovich transformation used in Problem 2.3, for example, is related to the last expression by a simple rotation $\chi(x) \to i\chi(x)$. The great utility of the transformation lies in the fact that it can be used to transform the action quartic in some field into a quadratic action, at the cost of introducing an additional variable. This then allows for a systematic formulation of the mean-field theory as the saddle-point of the integral over the Hubbard–Stratonovich field. The disadvantage of the method is that when there is more than one way to implement it, a particular Hubbard–Stratonovich transformation favors the selected type of ordering at the expense of other possibilities.

Appendix B

Linked-cluster theorem

Here we give a short proof of the linked-cluster theorem used in Section 3.2 to exclude the disconnected diagrams from the momentum-shell transformation. To be specific we assume a complex field $\Phi(\vec{r})$, although the proof is easily generalized to all other symmetries. Let us write

$$Z_<^p = 1 + p \ln Z_< + \sum_{m=2}^{\infty} \frac{(p \ln Z_<)^m}{m!},$$

where p is an arbitrary number, and $Z_< = Z/Z_{0>}$, with Z and $Z_{0>}$ as defined by Eqs. (3.12)–(3.15). The action for the slow modes $S_< = -\ln Z_<$ is therefore

$$S_< = -\lim_{p \to 0} \frac{\mathrm{d}Z^p}{\mathrm{d}p},$$

and equals the negative coefficient of the linear term in the expansion in powers of p.

On the other hand, for integer $p = n$ we may compute $Z_<^n$ by introducing n *replicas* of the complex fields:

$$Z_<^n = \frac{Z^n}{Z_{0>}^n}$$

with

$$Z^n = \int \prod_{i=1}^{n} \prod_{k<\Lambda} \frac{\mathrm{d}\Phi_i^*(\vec{k})\mathrm{d}\Phi_i(\vec{k})}{2\pi \mathrm{i}} \mathrm{e}^{-\sum_{i=1}^{n}(S_{0<,i}+S_{0>,i}+S_{int,i})},$$

with

$$S_{0<,i} = \int_0^{\Lambda/b} \frac{d\vec{k}}{(2\pi)^d} (k^2 - \mu) |\Phi_i(\vec{k})|^2,$$

$$S_{0>,i} = \int_{\Lambda/b}^{\Lambda} \frac{d\vec{k}}{(2\pi)^d} (k^2 - \mu) |\Phi_i(\vec{k})|^2,$$

and

$$S_{int,i} = \lambda \int_0^{\Lambda} \frac{d\vec{k}_1 \ldots d\vec{k}_4}{(2\pi)^{4d}} \delta(\vec{k}_1 + \vec{k}_2 - \vec{k}_3 - \vec{k}_4) \Phi_i^*(\vec{k}_4) \Phi_i^*(\vec{k}_3) \Phi_i(\vec{k}_2) \Phi_i(\vec{k}_1).$$

Similarly,

$$Z_{0>}^n = \int \prod_{i=1}^{n} \prod_{\Lambda/b < k < \Lambda} \frac{d\Phi_i^*(\vec{k}) d\Phi_i(\vec{k})}{2\pi i} e^{-\sum_{i=1}^{n} S_{0>,i}}.$$

$Z_<^n$ for a general n can now be computed perturbatively in the interaction λ, exactly as before, except that each field carries an additional replica index $i = 1, 2, \ldots, n$. Wick's theorem in Eqs. (3.21)–(3.23) still holds, and

$$\langle \Phi_i^*(\vec{k}_1) \Phi_j(\vec{k}_2) \rangle = (2\pi)^d \frac{\delta(\vec{k}_1 - \vec{k}_2)}{k_1^2 - \mu} \delta_{ij},$$

since $S_{0>}$ is diagonal in replica indices.

In the diagrammatic representation of the averages over the fast modes, as in Fig. 3.2 for example, each line carries a replica index, all the lines originating and terminating in the same interaction vertex have the same index, and the indices are summed from one to n. Each connected diagram therefore carries a single replica index, and after the summation yields a contribution to $Z_<^n$ proportional to n. A diagram with k connected pieces will therefore be proportional to n^k. Analytically continuing from integer to non-integer values of p and taking the limit $p \to 0$, we thus find that the action for the slow modes $S_<$ is determined only by the connected diagrams.

Appendix C

Gauge fixing for long-range order

We can eliminate gauge invariance in the Ginzburg–Landau theory in Eq. (4.2) by adding the so-called "gauge-fixing" term to L_{GL}:

$$L_{GL} \to L_{GL} + \frac{1}{2g}(\nabla \cdot \vec{A}(\vec{x}))^2.$$

The correlator of the vector potential at $e = 0$ is then finite and equal to

$$\langle A_i(\vec{q}) A_j(-\vec{q}) \rangle = \frac{T}{q^2}(\delta_{ij} + (g-1)\hat{q}_i \hat{q}_j),$$

as can be checked by inverting the three-by-three matrix in the wavevector space. Choosing $g = 0$ suppresses the fluctuations of the longitudinal component of the gauge field completely and the gauge-field correlator becomes transverse. If the gauge field is massive the result for $g = 0$ is just Eq. (4.10).

Let us now assume $T \ll T_c$ and write $\Phi(\vec{x}) = |\Phi_0| e^{i\phi(\vec{x})}$, where the amplitude $|\Phi_0|^2 = -a(T)/b$ is assumed constant. This should be justified at low temperatures where the fluctuations in the amplitude are strongly suppressed, and the phase is the only fluctuating degree of freedom. Up to a constant term, the Ginzburg–Landau theory for the remaining phase fluctuations at low temperatures may then be written as

$$L_{GL} = |\Phi_0|^2(\nabla\phi(\vec{x}) - e\vec{A}(\vec{x}))^2 + \frac{1}{2}(\nabla \times \vec{A}(\vec{x}))^2 + \frac{1}{2g}(\nabla \cdot \vec{A}(\vec{x}))^2.$$

Phase fluctuations are then evidently coupled only to the longitudinal component of the vector potential.[1] The integration over $\vec{A}$ therefore yields

$$L_{GL} \rightarrow |\Phi_0|^2 \frac{(\nabla^2 \phi(\vec{x}))^2}{-\nabla^2 + 2ge^2|\Phi_0|^2},$$

where the operator ∇ should as usual be understood as $i\vec{q}$ in the wavevector space. Note that for $g \neq 0$, upon performing the usual gradient expansion one finds that the coupling to fluctuating gauge field has made the spin-wave phase fluctuations at low wavevectors much less costly: $L_{GL} \sim (\nabla^2 \phi)^2$. For $g = 0$, on the other hand, spin-wave fluctuations of the phase become decoupled from the gauge field and the result is the same as if $e = 0$, i.e. $L_{GL} \sim (\nabla \phi)^2$.

The usual correlation function for the order parameter,

$$G_g(\vec{x}) = \langle \Phi^*(\vec{x})\Phi(0) \rangle \approx |\Phi_0|^2 \langle e^{i(\phi(\vec{x}) - \phi(0))} \rangle,$$

in the Gaussian approximation for spin-wave fluctuations at low temperatures becomes

$$G_g(\vec{x}) = |\Phi_0|^2 \exp\left[-\frac{T}{2} \int \frac{d\vec{q}}{(2\pi)^2} \frac{q^2 + 2ge^2|\Phi_0|^2}{|\Phi_0|^2 q^4}(1 - e^{i\vec{q}\cdot\vec{x}})\right],$$

and is evidently dependent on the gauge-fixing parameter g. If $g = 0$, $G_0(\vec{x})$ approaches a finite constant at large x, whereas if $g \neq 0$,

$$G_g(\vec{x}) = G_0(\vec{x})e^{-gTe^2|\vec{x}|I(\Lambda x)},$$

where

$$I(\Lambda x) = \int_0^{\Lambda x} \frac{d\vec{q}}{(2\pi)^3} \frac{1 - e^{i\vec{q}\cdot\vec{x}}}{q^4},$$

in three dimensions. Since the integral $I(\Lambda x)$ is convergent at large x, we find that for $g \neq 0$, $G_g(\vec{x})$ is exponentially small at large distances. Allowing any fluctuations of the longitudinal component of the vector potential therefore destroys the long range order in the standard correlation function $G_g(\vec{x})$ at any finite temperature.

A gauge-invariant correlation function can be constructed at the cost of having to average over a non-local object. Consider, for example,

$$G(\vec{x}) = \langle \Phi^*(\vec{x})e^{ie\int_0^{\vec{x}} \vec{A}_1(\vec{y})\cdot d\vec{y}} \Phi(0) \rangle,$$

---

[1] Vortex configurations for which this would not hold are excluded by our construction as quantitatively unimportant at low temperatures. Since they only additionally disorder the phase their inclusion would only strengthen our ultimate conclusion.

where the integral is taken along the straight line connecting the origin to the point $\vec{x}$. $\vec{A}_1(\vec{y})$ is the longitudinal component of the vector potential defined by

$$\vec{A}_1(\vec{q}) = (\vec{A}(\vec{q}) \cdot \hat{q})\hat{q}$$

in the wavevector space. Obviously, $G(\vec{x}) = G_0(\vec{x})$, and the gauge-invariant correlation function $G(\vec{x})$ coincides with the usual (gauge-dependent) correlation function if the latter is computed in the transverse gauge.

Bibliography

There exists a vast literature on the subject of phase transitions and critical phenomena. Here I provide a selection of pertinent books and articles that the reader may wish to further explore.

General reference

There are many good books on the theory of critical phenomena and renormalization group. A nice introduction may be found in N. Goldenfeld, *Lectures on Phase Transitions and the Renormalization Group* (Addison Wesley, 1992), and J. Cardy, *Scaling and Renormalization in Statistical Physics* (Cambridge University Press, 1996). A more technical, but a very pedagogical treatment is J. J. Binney, N. J. Dowrick, A. J. Fisher, and M. E. J. Newman, *The Theory of Critical Phenomena* (Oxford University Press, 1992). A thorough exposition with the emphasis on applications to "soft" condensed matter systems is provided by P. M. Chaikin and T. C. Lubensky, *Principles of Condensed Matter Physics* (Cambridge University Press, 1995).

Chapter 1

Interesting history of early scaling ideas and of the development of the renormalization group, together with the original references, may be found in the Nobel lecture by Kenneth Wilson in *Nobel Lectures in Physics*, edited by G. Ekspong (World Scientific, 1997).

The history of the Lenz–Ising model is presented by S. G. Brush, *Reviews of Modern Physics* **39**, 883 (1967). A discussion of Problem 1.3 may be found in R. Peierls, *More Surprises in Theoretical Physics* (Princeton University Press, 1991), Chapter 3.2. The rigorous version of Peierls' argument was constructed by R. B. Griffiths, *Physical Review* **136**, A437 (1964).

203

The critical exponents in Table 1 are as compiled in the book by J. J. Binney *et al.* cited above under General reference.

Chapter 2

A nice introduction to physics of ^{4}He may be found in J. Wilks, *The Properties of Liquid and Solid Helium* (Oxford University Press, 1967).

A detailed treatment of the coherent state representation of bosonic and fermionic partition functions is presented in J. Negele and H. Orland, *Quantum Many-Particle Systems* (Addison-Wesley, 1988).

A discussion of Landau's mean-field theory of phase transitions can be found in most books on phase transitions or statistical mechanics, including those cited under General reference. Still illuminating, Landau's original paper can be found in the *Collected Papers of L. D. Landau*, edited by D. ter Haar (Pergamon Press, 1965).

Quantum phase transitions, considered in Problem 2.5, are described in more detail in J. Hertz, *Physical Review B* **14**, 1165 (1976), M. P. A. Fisher, P. B. Weichman, G. Grinstein, and D. S. Fisher, *Physical Review B* **40**, 546 (1989), and I. F. Herbut, *Physical Review Letters* **85**, 1532 (2000).

Chapter 3

An excellent early review of the momentum-shell transformation and of the concomitant ϵ-expansion is K. G. Wilson and J. Kogut, *Physics Reports* **12C**, 75 (1974). A thorough exposition of the field-theoretic renormalization group may be found in the book by J. J. Binney *et al.* cited under General reference. High-order calculations and the resummation techniques are discussed, for example, in H. Kleinert and V. Schulte-Frohlinde, *Critical Properties of Φ^4 Theories* (World Scientific, 1997).

Critical behavior with more than one quartic coupling, treated in Problems 3.8–3.10, is discussed in A. Aharony, in *Phase Transitions and Critical Phenomena*, edited by C. Domb and M. Green, volume 6, 357 (Academic press, 1976), and more recently in P. Calabrese, A. Pelissetto, and E. Vicari, *Physical Review B* **67**, 054505 (2003).

Chapter 4

Ginzburg–Landau theory was introduced in the classic paper by V. Ginzburg and L. D. Landau, in *Collected Papers of L. D. Landau*, edited by D. ter Haar (Oxford University Press, 1965). A discussion of the dependence of the response of the superconductor to external magnetic field on κ may also be found in the same paper, and in many books on solid state physics. Microscopic theory of superconductivity in metals is presented in J. R. Schrieffer, *Theory of Superconductivity* (W. A. Benjamin Inc., 1964).

The application of the gauge theory to liquid crystals is discussed in P. G. de Genness and J. Prost, *The Physics of Liquid Crystals*, 2nd edition, (Clarendon Press, 1993).

Fluctuation-induced first-order transition in gauge theories was discovered by S. Coleman and E. Weinberg, *Physical Review D* **7**, 1988 (1973), and B. I. Halperin, T. C. Lubensky, and S.-K. Ma, *Physical Review Letters* **32**, 292 (1974).

Quantum phase transitions of itinerant electrons where the issue of coupling to low-energy excitations also naturally arises are discussed in D. Belitz, T. R. Kirkpatrick, and T. Vojta, *Reviews of Modern Physics* **77**, 579 (2005).

The effect of the magnetic field on the transition in liquid crystals examined in Problem 4.3 was studied in A. Yethiraj, R. Mukhopadhyay, and J. Bechhoefer, *Physical Review E* **65**, 021702 (2002).

Next-order β-functions in Problem 4.6 were computed by S. Kolnberger and R. Folk, *Physical Review B* **41**, 4083 (1990), and used to determine the first order correction to N_c in I. F. Herbut and Z. Tešanović, *Physical Review Letters* **78**, 980 (1997).

The anomalous dimension of the vector field and the implications for scaling of the penetration depth were discussed in I. F. Herbut and Z. Tešanović, *Physical Review Letters* **76**, 4588 (1996), and studied by Monte Carlo simulations in P. Olsson and S. Teitel, *Physical Review Letters* **80**, 1964 (1998), and in A. K. Nguyen and A. Sudbo, *Physical Review B* **60**, 15307 (1999).

Critical behavior of the penetration depth near the Wilson–Fisher fixed point in high-temperature superconductors (Problem 4.7) was observed in S. Kamal *et al.*, *Physical Review Letters* **73**, 1845 (1994).

The effects of order-parameter fluctuations on the first-order transition in type-I materials, considered in Problems 4.9–4.11, were studied in I. F. Herbut, A. Yethiraj, and J. Bechhoefer, *Europhysics Letters* **55**, 317 (2001).

Chapter 5

For a detailed discussion of the Nambu–Goldstone theorem in particle physics, see S. Weinberg, *The Quantum Theory of Fields*, Volume II (Cambridge University Press, 1996). The Mermin–Wagner–Hohenberg theorem was derived in D. Mermin and H. Wagner, *Physical Review Letters* **17**, 1133 (1966), and P. Hohenberg, *Physical Review* **158**, 383 (1967).

The non-linear σ-model was introduced by A. Polyakov, *Physics Letters B* **59**, 79 (1975). A field-theoretic analysis of the model was presented in E. Brézin and J. Zinn-Justin, *Physical Review B* **14**, 3110 (1976). For a recent discussion of possible causes of discrepancies with the $4 - \epsilon$ expansion, see E. Brezin and S. Hikami, preprint http://arxiv.org/abs/cond-mat/9612016. The non-linear σ-model representation of the wavefunction localization problem in random potential is discussed in A. J. McKane and M. Stone, *Annals of Physics (New York)* **131**, 36 (1981).

Higher-order calculations and the Borel–Páde approximation for the exponents are discussed in W. Bernventher and F. J. Wegner, *Physical Review Letters* **57**, 1383 (1986), and H. Kleinert, *Physics Letters A* **264**, 357 (2000).

An application of the scaling of the correlation length in Problem 5.2 to the antiferromagnetic state of high-temperature superconductors was considered in S. Chakravarty, B. I. Halperin, and D. R. Nelson, *Physical Review B* **39**, 2344 (1989).

Problems 5.3 and 5.4 are motivated by the treatment in D. R. Nelson and R. Pelcovits, *Physical Review B* **16**, 2191 (1977).

Problem 5.5 follows closely the analysis of J. M. Kosterlitz, *Physical Review Letters* **37**, 1577 (1976).

Chapter 6

The importance of vortices in the two dimensional XY model was pointed out by V. L. Berezinsky, *Soviet Physics JETP* **34**, 610 (1971). The mean-field theory of the Kosterliz–Thouless transition was formulated in J. M. Kosterlitz and D. J. Thouless, *Journal of Physics C* **6**, 1181 (1973).

The three dimensional plasma of logarithmic charges in Problem 6.2 was studied in I. F. Herbut and B. Seradjeh, *Physical Review Letters* **91**, 171601 (2003).

The Villain transformation was introduced by J. Villain, *J. Phys. (Paris)* **36**, 581 (1975). The duality transformation in two dimensions was developed in J. V. José, L. P. Kadanoff, S. Kirkpatrick, and D. R. Nelson, *Physical Review B* **16**, 1217 (1977).

The renormalization group transformation for the Coulomb plasma was formulated by J. M. Kosterlitz, *Journal of Physics C* **7**, 1046 (1974). The universal jump in superfluid density was derived by D. R. Nelson and J. M. Kosterlitz, *Physical Review Letters* **39**, 1201 (1977).

A pedagogical introduction to homotopy theory and its applications to physics of defects in condensed matter can be found in N. D. Mermin, *Reviews of Modern Physics* **51**, 591 (1979).

The quantum superfluid–insulator transition in one dimension in Problem 6.7 was considered by I. F. Herbut, *Physical Review B* **58**, 971 (1998).

The $2 + \epsilon$ expansion in Problem 6.10 was formulated in J. L. Cardy and H. W. Hamber, *Physical Review Letters* **45**, 499 (1980).

Chapter 7

A nice discussion of duality between the three dimensional XY model and the frozen lattice superconductor may be found in M. Peskin, *Annals of Physics (New York)* **113**, 122 (1978). The inverted XY behavior of the three dimensional lattice superconductor was demonstrated in C. Dasgupta and B. I. Halperin, *Physical Review Letters* **47**, 1556 (1981).

Compact electrodynamics was introduced by K. Wilson, *Physical Review D* **10**, 2445 (1974). The three dimensional compact electrodynamics in Problem 7.5 was studied by A. M. Polyakov, *Nuclear Physics B* **120**, 429 (1977).

The compact lattice superconductor in Problem 7.6 was studied by M. B. Einhorn and R. Savit, *Physical Review D* **17**, 2583 (1978); **19**, 1198 (1979). A different argument for the phase diagram in Figure 7.4 in Problem 7.7 can be found in N. Nagaosa and P. A. Lee, *Physical Review B* **61**, 9166 (2000).

Chapter 8

A comprehensive treatment of quantum phase transition can be found in S. Sachdev, *Quantum Phase Transitions* (Cambridge University Press, 1999). A nice introduction into finite temperature scaling, with applications to quantum Hall systems, is S. L. Sondhi, S. M. Girvin, J. P. Carini, and D. Shahar, *Reviews of Modern Physics* **69**, 315 (1997).

An early qualitative discussion of the Bose–Hubbard model was given in M. P. A. Fisher, P. B. Weichman, G. Grinstein, and D. S. Fisher, *Physical Review B* **40**, 546 (1989). Detailed computation of the coefficients in the expansion in Section 8.3, including the quartic coefficient, is presented in R. A. Lehrer and D. R. Nelson, *Physical Review B* **58**, 12385 (1998). A systematic strong-coupling expansion was performed in J. K. Freericks and H. Monien, *Physical Review B* **53**, 2691 (1996).

Universal conductivity in two dimensions was discussed, for example, in M. C. Cha, M. P. A. Fisher, S. M. Girvin, M. Wallin, and A. P. Young, *Physical Review B* **44**, 6883 (1991); A. van Otterlo, K.-H. Wagenblast, R. Fazio, and G. Schon, *Physical Review B* **48**, 3316 (1993); and K. Damle and S. Sachdev, *Physical Review B* **56**, 8714 (1997).

The effects of Coulomb interaction on the quantum superfluid transition treated in Problem 8.10 were considered in M. P. A. Fisher and G. Grinstein, *Physical Review Letters* **60**, 208 (1988); J. Ye, *Physical Review B* **58**, 9450 (1998); and I. F. Herbut, *Physical Review Letters* **87**, 137004 (2001).

Index